Australian Native Plants

Australian Native Plants

The Kings Park Experience

Editor: Mark Webb

PUBLISHING

National Library of Australia Cataloguing-in-Publication entry

Australian native plants : the Kings Park experience / edited by Mark Webb.

9780643103214 (pbk.)
9780643106994 (epdf)
9780643107007 (epub)

Includes bibliographical references and index.

Endemic plants – Australia.
Urban parks – Western Australia – Perth.
Botanical gardens – Western Australia – Perth.
Kings Park (Perth, W.A.)

Webb, Mark, 1957–, editor.
Botanic Gardens and Parks Authority (W.A.)

580.994

Published by
CSIRO PUBLISHING
369 Gardiner Road, Clayton VIC 3168
Private Bag 10, Clayton South VIC 3169
Australia

Telephone: [+613] 9545 8555
Local call: 1300 788 000 (Australia only)
Fax: +61 3 9662 7555
Email: csiropublishing@csiro.au
Web site: www.publishing.csiro.au

Mar26_RP_ILS

Front cover: photographs by David Blumer (top), Rosalie Okley (bottom)
Back cover: photographs by Rosalie Okley

Set in Minion Pro 10/15 and Helvetica Neue
Cover and text design by Andrew Weatherill
Typeset by Andrew Weatherill
Printed by Ingram Lightning Source

Contents

Contemporary plant display around Zamia café in Kings Park

David Blumer

Foreword

Gardening provides a powerful connection between people and plants worldwide. The choice of plants grown is a celebration of each gardener's aesthetic and practical connection to the world of plant diversity.

At a time of unprecedented environmental change and ongoing loss of biodiversity, the importance of gardening has assumed new dimensions. Each gardener can make a positive contribution – in a small or large way – to helping care for the diversity that underpins their enthusiasm for green companions in domestic and civic life. Indeed, it's becoming clear that without active help from gardeners, many more plants than the huge number already at risk would become threatened with extinction.

In a world with an estimated 400 000 species of flowering plants, and many more ferns, mosses and allies, there's plenty to choose from. Yet a fifth of the world's plant species – some 80 000 – are threatened with extinction based on recent estimates from leading botanical institutions, and a third are so poorly known that a reliable assessment of their conservation status is not possible on present information.

In this context, one strategy of inestimable value is for gardeners to grow plants local to their home country or district. Including just one or two local native plants in the horticultural palette is valuable in itself. Establishing whole gardens with such a selection is a new world entirely, full of novel aesthetic delights and enriching discoveries.

This book provides a compelling account of contemporary approaches to the horticulture of Australian plants. While its focus is on the not inconsiderable contributions made by Kings Park and Botanic Garden staff to the topic at hand, the book's contents have wider applicability, offering insights and approaches of national and international relevance.

Exciting discoveries and breakthroughs continue to be made in our time. The application of smoke as a stimulant for seed germination, celebrated herein, is just one such example. The great detective story in isolating the compounds responsible is summarised, illustrating the best of modern botanical and biochemical science.

But horticulture of Australian plants offers the potential for the humble home gardener to make significant contributions as well. With more than 25 000 species of flowering plants on the continent, 90% found nowhere else on Earth, and more than 10% yet to be named by botanists, let alone grown and understood, the Australian flora is one great field of everlasting discovery and enjoyment.

What is the name of that species that's caught your eye? How do you collect seed, responsibly? How is a species of interest best propagated? What management does it need to bring it into horticulture successfully? How can gardeners gain better access to the rich diversity of Australian plants suitable for horticulture? What improvements in techniques are really making breakthroughs? How can gardening be undertaken in a way that enhances the conservation of plants and animals?

All this and more is addressed in this richly illustrated and well-produced volume. I encourage readers to explore its contents, savour its visual delights, muse on its practical suggestions, and help celebrate and conserve one of the richest floral heritages on the planet through active, informed gardening.

There is a fundamental message of hope in growing plants and sharing your life with them. There never was a more important time to know and grow plant diversity. Herein lies the key to making a start, or honing your skills if you're already some way down the path, and share a passion for Australian plants as the authors of the book so evidently do.

Happy gardening!

Professor Stephen Hopper AC
Director (CEO and Chief Scientist)
The Royal Botanic Gardens, Kew

Acknowledgements

I wish to express sincere thanks to my colleagues who authored or co-authored chapters in this book: Grady Brand, Dr Eric Bunn, Digby Growns, Dr David Merritt, Amanda Shade, Luke Sweedman and Jeremy Thomas from the Botanic Gardens and Parks Authority (BGPA); Dr Elaine Davison from Curtin University (formerly of the Department of Conservation and Land Management and the Department of Environment and Conservation Western Australia); and Aileen Reid, Dr Kevin Seaton and Bill Woods from the Department of Agriculture and Food, Western Australia (DAFWA). A special thanks to David Blumer from the BGPA who, in addition to being an expert horticulturist, is also a keen photographer and provided most of the photographs in the Introduction and Chapters 1, 2, 3 and 5. The diagrams in Chapter 5 were expertly done by Rosalie Okley and the initial editing for all chapters was done in association with Julia Berney. Thank you to Nick Alexander and Tracey Millen from CSIRO PUBLISHING for their patience and support.

While I have been responsible for coordinating this book, it has really been the enthusiasm of each of the chapter authors and co-authors that transformed an idea into a reality. A final note of thanks to our many other colleagues, past and present, who have helped to shape the collective Kings Park experience in growing Australian native plants and who freely shared their knowledge and experience with us.

Mark Webb
Chief Executive Officer
Botanic Gardens and Parks Authority

Introduction

Mark Webb

For the most part, Australia is an arid or semi-arid continent with warm to hot summers and mild winters. Rainfall can be highly variable and occurs mainly in summer in the north and in winter across the south. Australia has many different soil types with most being very nutrient deficient. It is also home to about 10 per cent of the world's flowering plants, which are increasingly being used in public landscapes and home gardens.

BGPA

Lechenaultia biloba is an attractive, small herbaceous plant from the south-west of Western Australia. It is now grown in public landscapes and home gardens.

Early private and public gardens in Australia generally focused on the use of non-native or exotic plants. However, in the late 19th century and early 20th century there was increasing interest in the collection and display of Australian native plants by some public and private gardens including the Royal Tasmanian Botanical Gardens, the Royal Botanic Garden Sydney, Maranoa Gardens in Victoria and Wittunga Garden in South Australia. But this interest did not spread to the majority of home-owners or those responsible for planning amenity landscapes.

Melbourne critic and social commentator AA Phillips coined the term 'cultural cringe' after the Second World War. It was mainly used in connection with the arts and contended that Australians were more likely to dismiss their own culture as inferior to the cultures of other countries, especially Great Britain and Europe. This assessment arguably also applies to Australians' love, or otherwise, of our natural landscapes. Many people, raised in countries with a more temperate climate and lush landscapes, neither understood nor felt at ease with the Australian bush, which can appear to be harsh, prickly and 'grey'.

Dating from the middle of the last century, most homes in Australia had space for a lawn at the front and back, and a garden filled with mainly exotic plants around the perimeter of the block. This style of gardening is what many people became used to and felt most comfortable with.

The late George Seddon, an academic in many disciplines and one of Australia's most respected environmental scholars, believed that the yearning for a lush garden did not sit well in many parts of Australia that experienced a generally harsh and dry climate. Plant selection and garden design, he believed, should be sympathetic to the local environment.

Native plant gardens were the tragic gardening story of the 1970s in Australia. In an attempt to connect with the natural landscape, use less water or reduce maintenance requirements, many home gardeners and landscapers experimented with a relatively small range of Australian plants. However, they didn't realise that without proper plant selection and judicious management, Australian plants can become unruly and the gardens overgrown and untidy. This experiment didn't deliver the style of garden that people felt comfortable with, or that added value to their outdoor living and it effectively slowed for a couple of decades a wider interest in growing Australian plants.

The nursery industry produces plants in response to public demand and, except for native plant enthusiasts, most Australians have traditionally purchased exotic plants.

David Blumer

Native plants in their natural habitat in Western Australia.

This lack of demand for native plants, together with various difficulties in their propagation and production requirements, contributed to a relatively limited suite of Australian plants being available for public landscapes and home gardens. Until the late 1990s, the supply of a wider range of Australian plants was restricted to specialist nurseries. However, since then and in response to demand especially from the landscaping industry, mainstream nurseries are now providing native plants in increasing numbers. As more information becomes available about production techniques and growing requirements, the range of Australian plant species available is also increasing.

With unreliable or decreasing rainfall and limited water supplies across many areas of southern Australia, and advocacy from presenters of television and radio programs, interest in growing Australian native plants has burgeoned. This interest has coincided with a trend towards larger houses on smaller blocks that has meant home-owners are becoming more purposeful about the style of garden they want and how they will interact with it.

Water-wise and low maintenance landscaping with Australian native plants is about appropriate design and using

David Blumer

Cultivated Australian plants on display in Kings Park, Perth.

plants that are relatively drought tolerant. Australian plants and especially those indigenous to a region deliver many environmental and ecological benefits. However, there are plants from other areas of Australia or overseas that are also low water users and provide a range of flower colours, variation in foliage type and plant forms suitable for public landscapes or home gardens. Complementary plantings where easy-to-grow local plants are combined with selected non-local or exotic plants can be a useful transition for people who are interested in having an attractive, functional garden but also want to grow more Australian plants.

David Blumer

A courtyard planting with Australian annuals and perennials.

For more than 50 years, information on growing Australian plants from groups such as the Australian Native Plants Society and publications from experts such as Ken Newby (*West Australian Plants for Horticulture* 1958: Parts 1 and 2) have contributed to the knowledge of Australian plants. Over the past 30 years, information on growing Australian plants has become more readily available through other publications and promotion from experts including Ivan Holliday, John Wrigley, Murray Fagg, Rodger and Gwen Elliot, David Jones, John Colwill, George Lullfitz, Jackie and Alec Hooper, Don Burke, Angus Stewart, Sabrina Hahn, Sue McDougall, Neville Passmore and Trevor Cochrane.

Australia's botanic gardens have long been a source of information about growing both exotic and Australian plants. The knowledge and expertise of garden staff has resulted in the display of many Australian plants and the transfer of valuable information to professional horticulturists and home gardeners alike.

Located in Perth, Western Australia, just a kilometre from the city's Central Business District, Kings Park and Botanic Garden is renowned for its displays of Australian plants. Although most of Western Australia's 12 000 native plants flower from July to October, the rich diversity of species means that, at any time of year, the many millions of visitors to the park can enjoy the constantly changing floral displays or the

BGPA

Kings Park and Botanic Garden in Perth, Western Australia.

A tree-lined avenue of *Corymbia citriodora*, native to eastern Australia, in Kings Park.

architectural grandeur of tree-lined avenues and individual tree specimens.

When a new botanic garden was opened in Kings Park in 1965 to grow mainly Western Australian native plants, almost nothing was known about their horticulture. In 1970 Arthur Fairall, the Kings Park superintendent, wrote *Western Australian Native Plants in Cultivation*. This book, now out of print, was based on the author's knowledge of local plants observed in their natural habitat and grown in Kings Park. It provided cultural information on more than 600 Western Australian plant species and was an important reference for many nurserymen and gardeners about which local native plants to grow and how to grow them.

Plant hardiness maps have been developed for key growing regions around the world. The Australian hardiness map produced by the Australian National Botanic Gardens locates Kings Park in Zone 4 with characteristic warm to hot summers with low humidity, and cool, moist winters. This map, together with maps that have been produced for other world hardiness zones, assists in identifying the climatic range for selected plants and can be used to extrapolate suitable locations for growing local plants in other areas of Australia and overseas.

Through investigation and practice, staff at Kings Park have demonstrated that Australian plants can be used to complement traditional and contemporary architecture. Their adaptability to a range of environments, and capacity to deliver year-round flowering displays means Australian plants are suitable for large landscapes and small gardens alike.

There are many books and other publications on growing and displaying Australian plants (see the 'Further reading' list). Television and radio programs, online information and forums have also added to the wealth of knowledge about the Australian flora. This book tells of the experience of Kings Park staff in the horticulture of mainly Western Australian native plants by providing the more technically minded professional or enthusiast with information based on decades of research, experimentation and application. From describing the growing conditions for Australian native plants to identifying the top

David Blumer

Border display using Australian plants.

ornamental plants and explaining how to propagate, select and breed them, the information in this book aims to encourage the growing of Australian plants so that they can be used more widely and contribute to interesting, attractive and diverse public landscapes and home gardens in a changing environment.

Mark Webb
Chief Executive Officer

Chapter 1

Growing Australian native plants

Mark Webb and Grady Brand

Australian plants need normal horticultural care and attention for optimum results. To achieve good plant growth, desirable form and strong flowering, many species require regular irrigation, pruning and applied fertiliser. Others, however, especially those in the Proteaceae family such as grevilleas and hakeas, require minimal ongoing care and attention once established.

Swainsona formosa growing in Kings Park.

Most Western Australian native plants grow best on well-drained, slightly acid soils and in locations that experience warm, dry summers and cool, moist winters. They prefer a sunny aspect and soils free of disease, especially the fungal diseases phytophthora and armillaria.

Plants that are indigenous to an area usually give the best results in that location. They also have the advantage of being very attractive to local wildlife. However, many Australian plants grow well in a range of environments and most wildlife adapts readily to introduced plants.

Plants from different regions of Western Australia growing under the same environmental and horticultural conditions in Kings Park.

When to plant

Australian native plants can be planted at any time of year provided they receive adequate water during warm to hot weather. The best time to plant in areas with winter rainfall is from early autumn to late spring. In areas that experience prolonged periods of cold weather or light frosts, it is preferable to plant after mid-spring.

Healthy seedlings ready for planting.

Planting

Generally, the younger the seedling to be planted, the better its establishment will be, provided it is hardened off and has a well-developed root system. If plants are held in pots for too long before being transplanted, they can become root-bound and may not grow to full size. Root-bound plants can also die unexpectedly up to several years after planting, especially following periods of prolonged hot weather. Purchasing plants from a reliable and accredited supplier will help to ensure a long-lived and healthy plant.

For large shrubs and trees, best long-term results are achieved by seeding directly into a three-dimensional air root-pruning pot (25 to 50 litres) and then planting out when the shrub or tree reaches a suitable size.

A healthy eucalypt in a three-dimensional air root-pruning pot ready for planting.

Make the planting hole at least 50 per cent wider and deeper than the pot; remove the seedling or plant from the pot, loosen the root ball before planting and then backfill with local soil. Make sure the stem of the plant is at the same level in the soil as it was in the pot, and water the plant into its new position.

In heavy or very compacted soils, break up the soil and only plant species that will tolerate such conditions. In these and other difficult situations, mounding with a free-draining sandy soil is a proven technique for successfully growing a wider range of Australian plants. The mound should be 30 to 50 centimetres above the existing soil line.

It is important to investigate and document any previous use of the site that has impacted on the soil, such as compaction from vehicles or high levels of residual nutrients (especially phosphorus) from other horticultural activities. These impacts can negatively affect plant growth and need to be factored into site design and plant selection.

Fertiliser

Most Australian native plants grow naturally in soils with low levels of residual phosphorus. When purchasing a complete fertiliser that contains a range of nutrients, choose one that has low or nil phosphorus and apply it according to the instructions on the label.

Controlled or slow release synthetic fertilisers are usually more expensive than fast release fertilisers per unit of available nutrient, but they are easy to use and more forgiving if accidentally over-applied.

Applying slow release fertiliser.

Fresh animal manures are not recommended for most Australian plants. Some, especially fresh chicken manure, can damage plants and also contain high levels of available phosphorus. Composted organic matter low in phosphorus can be used on Australian plants and has the added value of supplying organic material that provides many benefits to the physical, chemical and microbiological environment of the soil.

Using fast release fertilisers requires experience and care. If excessively applied on young seedlings or even on older plants, fast release fertilisers (especially those that contain higher concentrations of nitrogen or potassium salts) can burn the roots causing leaf wilting, leaf burning and even plant death. Always apply any fertiliser according to the instructions on the label.

If the plant has been bought from a reliable supplier, there should be enough slow release fertiliser in the potting mix for the plant to thrive for several weeks after planting before additional fertiliser has to be applied.

Most Australian plants don't need added fertiliser for maintenance growth, but for many species, application of a slow release or complete fertiliser with low or nil phosphorus in late spring to early summer, or after flowering, will improve plant vigour and flowering display. Plants grown in pots will usually require more regular fertilising than plants grown in the ground.

Many Australian plants are adapted to growing in nutrient-deficient soils and some have developed mechanisms such as proteoid roots (clusters of roots near the soil surface) to extract soil phosphorus at very low concentrations. The regular application of fertiliser containing phosphorus, or growing plants on land previously fertilised for other crops, can result in too much phosphorus being available to some plant species. Phosphorus toxicity is well known in banksias but it can also occur in other genera in the Proteaceae, as well as genera in the Rutaceae, Fabaceae, Mimosaceae, Myrtaceae and Haemodoraceae. Phosphorus toxicity symptoms vary but can include slow growth, yellowing of the youngest leaves (a symptom which may be confused with iron deficiency), burning of the leaf edges, reddening and then death of the older leaves. However, for most banksias that suddenly yellow and die, disease is usually the more likely cause, not nutrient toxicity.

BGPA

Proteoid or cluster roots on *Banksia attenuata*.

For Australian plants that grow naturally on acid soils, growing them in alkaline soils or irrigating with alkaline water can be very challenging. Lime-induced chlorosis or iron deficiency typically shows up as a yellowing of the youngest leaves and may be corrected in the short term by application of iron chelate. Foliar applications are quicker acting than soil applications but the effect may only last a few weeks or months. Slowly changing the soil pH to accommodate the more acid-loving plants can be done but is not a straightforward or quick process. Using a less sensitive cultivar of the same species can sometimes overcome iron deficiency problems, but if the symptoms persist, grow the plant in a pot using a high quality potting mix, on an alkaline-tolerant rootstock if one is available, or try another native plant better suited to alkaline conditions.

Watering

Australian native plants have varying water requirements and an understanding of their natural environment is a good guide as to how they will perform in a cultivated situation. Many plants in the Proteaceae, such as grevilleas or banksias with their proteoid or cluster root systems, usually need supplementary watering only during establishment and in very hot weather.

Others, including boronias such as *Boronia megastigma* and *B. heterophylla* from the cooler south-west forest regions of Western Australia, require regular watering to survive and thrive. These plants die within a few days if water-stressed and are commonly known as 'drop-dead' plants. While they can be grown in-ground, if irrigation is likely to be irregular or limited, then growing them in self-watering pots can give better medium term results. In contrast, *B. crenulata*, also from the south-western region of Western Australia, is relatively hardy and can survive in-ground with less regular watering.

Increasingly, grafting site-specific plants onto more vigorous or tolerant rootstocks such as *B. megastigma* on *B. clavata* is extending the range of locations in which more difficult to grow or sensitive species can be grown.

After planting, keep the plants well watered, but not waterlogged, until they have established. Be especially diligent in applying enough water in early autumn and late spring, as plants can quickly dry out after only a few days of unseasonably hot weather, especially on well-drained soils.

Plants with similar watering requirements should be grouped together for ease of management, although the need to do this can be overcome to some degree if trickle irrigation, with different output drippers or mini-sprinklers on the same line, is used on plants with varying water requirements. On difficult-to-wet or non-wetting soils, regularly applying a wetting agent and irrigating at a slower rate will improve water penetration.

David Blumer

An example of sub-surface trickle irrigation (the soil has been removed).

In Kings Park, permanent display beds with excellent plant growth and flowering have been achieved by using specialised sub-surface irrigation. In these situations, the drip line is located 5 centimetres underground, with in-line drippers at 30 centimetres and in rows 0.5 metre to 1 metre apart, with the soil backfilled. Once plants have been established on this system, it is only necessary to irrigate once or twice per week to a depth of 30 centimetres each time in the warmer, drier months with no additional water required in the cooler, wetter months.

There is an opportunity with many Australian native plants to establish a low- or nil-water-use garden. This trend is becoming increasingly popular as water supplies become more limited and the cost of water rises. Importantly, low water use gardens using indigenous plants represent a more contemporary and sustainable connection with the Australian landscape. However, there are Australian plants such as some boronias and other herbaceous plants that are not suitable for low input gardens, as they require more water and regular maintenance for optimum results.

If plants are started on supplementary watering with the intention of eventually moving to a low- or nil-water-use garden, slowly reduce the quantity and frequency of the watering over several months to acclimatise the plants to a changed moisture regime.

David Blumer

An Australian plant garden managed by Kings Park Volunteer Master Gardeners. It is watered only by rainfall and receives no supplementary irrigation.

Mulching

Mulching is one of the best horticultural practices to improve any public landscape or home garden and should be used even when growing plants in pots. A layer of mulch about 5 centimetres thick will make the bed more attractive, reduce weeds, aid earthworm and microbial activity, keep the roots cooler and conserve soil moisture. Suitable organic mulches include screened pine bark, eucalyptus chips, composted organic matter and shredded leaves. As the mulch decomposes, new mulch should be added and this process will significantly improve the soil over time. Keep the mulch clear of the plant stem as moist or wet mulch in contact with it can lead to stem cankers and rots.

David Blumer

Mulching newly planted garden beds in Kings Park.

Pruning

Most Australian plants respond to pruning. It gives the plants shape, encourages vigorous and healthy new growth, and improves flowering displays. Most Australian plants flower from midwinter to late spring and the best time to prune is after flowering has finished. There are essentially three types of pruning: formative, maintenance and timing-related.

Formative pruning involves a light tip-pruning of herbaceous plants and shrubs, usually in the first 12 to 18 months after planting, to shape the plant and encourage it to send up multiple shoots.

David Blumer

Pruning to remove about 30 per cent of mature plant after flowering.

Ongoing maintenance pruning involves a light to moderate tip-pruning when flowering is finished, to help rejuvenate the plant for the next year. Some plants such as Geraldton wax, pimeleas, boronias, verticordias, eremophilas and grevilleas respond well with up to 30 per cent of their foliage being pruned back each year after flowering. With *Grevillea* species and hybrids that flower for many months, pruning in spring ensures that there will be good flowering the next year. Only prune into wood with green leaves below the cut until the response by the plant to pruning is understood – some plants don't shoot from old wood when actively growing leaves are not present.

The third type of pruning involves timing. For example, on *Banksia coccinea* the flowers for the next season are being set before the current season's flowering has finished. In this situation, regularly removing just the spent flowers over the last few weeks of this season's flowering will ensure a reasonable flowering for the next year, with judicious maintenance pruning then done as required. In contrast, for *B. baxteri* and most popular Australian plants, the flowers for next season are set well after this year's flowering, so normal pruning straight after flowering is suitable.

Lesley Hammersley

Scaevola 'Blue Print' in a hanging basket.

Ten tips for successfully growing Australian native plants

- Choose the right plant for your climate and soil type.
- Buy healthy plants from a reliable, accredited supplier.
- If the plant does not grow in your situation, get rid of it; there are plenty of others to try.
- Try growing Australian plants in pots.
- Prune shrubs and herbaceous plants after flowering.
- Treat herbaceous perennials such as scaevolas and most kangaroo paws as annuals or biennials to get the best displays.
- Mulching works; do it.
- Australian plants respond well to fertiliser – just don't overdo it.
- Don't apply phosphorus fertiliser unless stated otherwise on the plant label.
- Don't believe everything you hear or read about growing Australian plants. There are exceptions to almost every rule, so be brave and experiment!

Weediness

One important factor to consider when growing any plant outside of its natural environment is that some species or cultivars may have the potential for weediness. Consider, therefore, the location in which a new plant will be cultivated, particularly if there are areas of natural vegetation nearby. For example, the coastal form of *Lobelia anceps*, a species native to the south-west of Western Australia, is no longer cultivated in Kings Park due to its potential to become a weed in the park's display garden beds and thus spread into nearby bushland. Other Australian plant species that have weediness tendencies in Kings Park include *Eucalyptus camaldulensis*, *Agonis flexuosa* and *Callitris preissii*. A well known example of weediness is *Melaleuca quinquenervia*, native to the east coast of Australia, New Caledonia and New Guinea. It is a serious invasive weed in the Everglades in Florida, USA, where it forms very dense stands, interrupting natural water flows and promoting extremely hot crown fires.

Fortunately, very few plants exhibit this tendency but it is a factor that needs to be considered when introducing any new plant. Traditionally, botanic gardens world-wide test many new plant cultivars and, together with the local nursery industry association or agriculture department, identify the potential for a particular species to become weedy.

Chapter 2

Groundcovers and shrubs

Grady Brand

Combining groundcovers with small and medium sized shrubs provides detail and complexity to a landscape and delivers high impact floral displays. Most groundcovers are relatively short-lived and are usually more heavily planted in areas where an intensive visual outcome is desired, with skilled horticulturists maintaining the display.

You should have a replacement strategy for groundcovers because most of the desirable and attractive species, such as *Scaevola aemula* 'Purple Fanfare', may require annual replacement while others such as *Leucophyta brownii* and *Kennedia prostrata* benefit from supplementary planting each year to ensure the garden maintains its vigour and high quality presentation.

Shrubs deliver the best display results when there is a range of ages in the planting. It is usually only necessary to replace shrubs if they become too woody or too old, or when a different display effect is required. The art of successfully maintaining a dynamic garden is in ensuring that the landscape contains a range of different aged plantings, with removal and replacement as part of a deliberate strategy. Annual pruning of shrubs after flowering will help maintain the desired shape and promote reliable flowering.

For best results, especially in larger areas, mass plant or 'drift' plant groundcovers and shrubs throughout the site, mixing species of different heights, forms, flowering time and colour. Annuals can be used to augment the displays.

The use of iconic species such as *Xanthorrhoea preissii* and *Macrozamia fraseri* creates an instant effect. These signature species denote an Australian identity within a landscape. They are best displayed above a suite of lower growing species that draw the eye to their large, architectural forms.

The following selection of groundcovers and small to medium shrubs is based on experience gained under cultivation in the Western Australian Botanic Garden in Kings Park. Proven performers, they are used as a focus to celebrate the diverse flora of Western Australia and to illustrate this diversity for the many visitors to Kings Park. All the selected plants are perfect for a low maintenance garden and, unless stated otherwise in the individual descriptions, are suitable for pot culture and will attract birds.

Prostrate banksia (*Banksia blechnifolia*)

Prostrate banksia is a groundcover growing to 0.7 metres high and 2 to 3 metres wide. The reddish new foliage and fern-like leaves make this a most attractive plant. One of the spring flowering banksias, it comes in a range of colours including red, pink, cream, orange and brown, with velvety flower spikes of up to 16 centimetres in length. Mass planting will achieve a strong landscape impact and is ideal for rockeries and borders, or in an elevated setting where the flowers are displayed over a low retaining wall.

Amanda Shade

Prostrate banksia (*Banksia blechnifolia*).

Honeypot dryandra (*Banksia nivea*, formerly *Dryandra nivea*)

Honeypot dryandra is a compact shrub that grows to about 1 metre by 1 metre. It has attractive, fern-like foliage. This species flowers over a long period, often bearing a few flowers throughout much of the year. Its main flowering time is during spring with flower colours including cream, yellow, orange, pink, red and brown. Generally the flower heads are up to 4 centimetres in diameter. The leaves are quite distinctive, creating visual contrast and impact within a mixed planting. Honeypot dryandra may be planted singly or *en masse*. It is well suited as an understorey shrub able to tolerate lower light situations, and ideally planted with other species that tolerate a semi-shade aspect such as *Chorizema cordatum*, *Adenanthos cunninghamii* and the prostrate form of *Dodonaea ceratocarpa*.

David Blumer

Prostrate banksia (*Banksia blechnifolia*).

Fanflower (*Scaevola aemula*)

David Blumer

Fanflower is a profusely flowering, quick-growing groundcover which grows to a height of 0.3 metres and a width of 0.6 metres. The five-petalled mauve to purple flowers are most abundant during the early summer months. However, it is possible to maintain flowering all year round provided that plant vigour is maintained. The well known cultivar *Scaevola aemula* 'Purple Fanfare' is relatively short-lived, with best results achieved in its first year, although it may live longer in optimum conditions. If vigour is lost, as tends to happen over the winter months, it is best to remove and replace plants after they have finished flowering. A new Kings Park hybrid, *Scaevola* 'Blue Print', is a more free-flowering cultivar and usually provides good displays for 2 or even 3 years after planting. Fanflower is most suitable for rockeries, hanging baskets and mass planting, or as a substitute for ornamental annuals. Planted in association with *Xerochrysum bracteatum*, dwarf *Anigozanthos* hybrids, *Leucophyta brownii* and *Conostylis candicans*, it gives excellent display outcomes.

David Blumer

Fanflower (*Scaevola aemula*).

Woolly bush (*Adenanthos sericeus*)

David Blumer

Woolly bush is a large, upright shrub that grows from 0.5 to 5 metres high by 2 metres wide. Its soft, silver-grey foliage provides an attractive contrast in the garden. It can be grown in a range of conditions, including coastal areas. The small red flowers are almost inconspicuous but they are a great source of nectar for small honeyeaters throughout most of the year. Woolly bush is best grown in full sun but is highly adaptable and can tolerate semi-shade to create an understorey feature. Full shade will cause plants to become leggy and more regular pruning is then required to maintain shape and form. This shrub is excellent as a windbreak, hedge or screen, or planted in association with *Scaevola crassifolia*, *Melaleuca huegelii* and *Conostylis candicans*.

David Blumer

Woolly bush (*Adenanthos sericeus*).

Geraldton wax (*Chamelaucium uncinatum*)

David Blumer

Geraldton wax is a fast-growing, large shrub that can reach a height and width of 4 metres, but smaller forms are available. It flowers profusely in late winter and spring with colours ranging from white through to deep purple. Individual flowers can be as large as 2.5 centimetres across.

Geraldton waxes are attractive when planted in a group or as individual specimens, or when used as a hedge. Planting a range of the different colour forms together can produce a great effect, especially when used on a larger scale for public display. Try mixing them with *Eucalyptus caesia*, *Grevillea thelemanniana* subsp. *glabrilimba* and any of the taller *Anigozanthos* hybrids. Geraldton wax is not suited to long-term pot culture.

David Blumer

Massed planting of hybrid kangaroo paws and small Geraldton wax (*Chamelaucium uncinatum*) with tree overstorey behind.

Burdett's banksia (*Banksia burdettii*)

David Blumer

This medium shrub has a compact habit and can grow to a height of 4 metres and a width of 2 metres. The two-toned orange and cream, cone-shaped flower spikes are spectacular against the blue summer sky. They can grow to 12 centimetres in length and 10 centimetres in diameter. Sometimes planted for use as a cut flower, this species flowers in summer and autumn, providing nectar-loving birds with a source of food during the hotter months. Mass planting of Burdett's banksia provides a bold statement within the landscape. Planting in association with *Eucalyptus lane-poolei*, *E. kingsmillii*, *E. kruseana*, *Melaleuca fulgens*, *Acacia aneura*, *Conostylis candicans* and *Banksia blechnifolia* is recommended.

David Blumer

Burdett's banksia (*Banksia burdettii*).

Waxflower (*Chamelaucium ciliatum*)

This is a medium shrub that grows to 1.2 metres high by 0.5 metres wide, and it has many forms with proven horticultural potential. It flowers profusely during spring. The small flowers of white, pink or purple make great fillers for floral arrangements. Ideally suited to mass planting, this species enjoys a full sun position but will adapt to semi-shaded positions. It is very effective when planted with other *Chamelaucium* species, and plants which have greater structural form, such as *Banksia ashbyi* subsp. *boreoscaia*, *B. baueri*, *B. pilostylis*, *Pimelea ferruginea*, *Eucalyptus caesia* and *E. victrix*.

David Blumer

Waxflower (*Chamelaucium ciliatum*).

Grevillea preissii subsp. *glabrilimba*

This compact, medium shrub grows to a height of 0.7 metres and a width of 1.2 metres. It has soft, green-grey foliage. The main flowering season is from July to September but under cultivation it can flower intermittently throughout the year. The bright red flowers attract small nectar-seeking birds. As a limestone lover this shrub is particularly suited to exposed coastal conditions. It requires a free-draining soil and is well suited to mass planting, making an ideal border. It can adapt to semi-shade positions as an understorey plant beneath an open canopy of tree species such as *Eucalyptus caesia*, *E. sepulcralis*, *E. lane-poolei* and *E. pyriformis*.

David Blumer

Grevillea preissii subsp. *glabrilimba*.

Cushion bush (*Leucophyta brownii*)

This species comes in a range of shapes, from the upright Canal Rocks form (1 metre high) to a dense, small cushion bush (0.15 metres). Its unique, almost white tomentose foliage and stems are a distinctive feature. The inconspicuous yellow globular flower heads appear in summer, but the plant is grown primarily for its foliage rather than its flowers. Cushion bush grows well under coastal conditions and is well-suited to being planted *en masse* as a border or as an accent plant to achieve varying degrees of grey foliage within a garden design. Planting in association with *Scaevola* 'Purple Fanfare', *Scaevola* 'Blue Print' and *Anigozanthos* 'Bush Pearl' or 'Bush Inferno' produces a complementary effect. Cushion bush does not attract birds.

David Blumer

Cushion bush (*Leucophyta brownii*).

Anigozanthos hybrid 'Big Red'

This strappy-leafed, herbaceous, perennial kangaroo paw is a hybrid between *Anigozanthos flavidus* and *A. rufus*. It grows to approximately 0.6 metres high with flower spikes reaching up to 2 metres. The large, bright red flowers are displayed at the ends of the long upright spikes in late spring and early summer. For maximum year-round display potential this hybrid (and any of the other tall kangaroo paw hybrids) is best planted in drifts in a mixed shrubbery or as accent planting. Avoid block planting because, when the flowers are spent, these plants lack sufficient structural integrity to maintain a presence in the landscape. There are many proven *Anigozanthos* hybrids available, with some of the best tall ones being 'Orange Cross', 'Pink Surprise', 'Yellow Gem', 'Bush Haze' and 'Bush Dawn'. Some of the best smaller growing cultivars include 'Bush Spirit', 'Bush Pearl' and 'Kings Park Federation Flame'.

David Blumer

Anigozanthos hybrid 'Big Red'.

Grey cottonhead (*Conostylis candicans*)

David Blumer

This grey-foliaged, strappy-leafed plant has a compact habit, 0.4 metres high and 0.4 metres wide. Bright yellow flowers appear in spring on terminal heads. The flowers provide a good contrast to the grey foliage. Grey cottonhead is able to tolerate coastal influences. For best effect plant *en masse* as a garden border or in association with a range of coastal-loving species such as *Scaevola crassifolia*, *Pimelea ferruginea*, *Templetonia retusa*, *Leucophyta brownii* and *Xanthorrhoea preissii*.

David Blumer

Grey cottonhead (*Conostylis candicans*).

Banksia pilostylis

This medium, dense shrub is multi-branched and grows to a height and width of 1.5 metres. *Banksia pilostylis* provides a striking floral display with profuse green to yellow cylindrical flower spikes up to 10 centimetres long by 5 centimetres wide. It flowers in spring and early summer. This plant provides a great accent at focal points and is also useful scattered throughout a mixed Australian shrubbery with plants such as *Anigozanthos flavidus*, *Chamelaucium uncinatum*, *Hakea orthorrhyncha*, *Melaleuca fulgens*, *Senna artemisioides* and *Eremophila glabra*.

David Blumer

Banksia pilostylis.

Banksia ashbyi subsp. *boreoscaia* (dwarf form)

The dwarf form of *Banksia ashbyi* is compact and low-growing, reaching a height and width of 1.5 metres. It has long, coarse leaves. This plant flowers spasmodically over an extended period, with the main period being winter to early spring. The cylindrical flowers are an intense orange and can be up to 15 centimetres in length and 10 centimetres in diameter. A great accent plant, this also looks good as a drift throughout a low shrubbery. Using coarse sand, aggregate or rocks accentuates the display. This subspecies is ideal for incorporating in a heathland style garden with species such as *Verticordia chrysantha*, *Grevillea preissii* subsp. *glabrilimba*, *Thryptomene denticulata*, *Banksia nivea* and *Hemiandra pungens*.

David Blumer

Banksia ashbyi subsp. *boreoscaia* (dwarf form).

Chenille honey myrtle (*Melaleuca huegelii*)

Chenille honey myrtle is a large, dense shrub growing to 4 metres by 4 metres. It has small, dark green, scale-like leaves. In late spring and early summer large, white, spike-like flowers appear on dense terminal heads. They are striking, very aromatic and very attractive to insects and nectar-loving birds. This species is a coastal lover; it is ideal for mass planting to produce a screen effect. It can be included in any coastal native plantings to provide a source of nectar when the main spring flowering of other species is over. It is very effective when planted in association with *Scaevola crassifolia*, *Conostylis candicans*, *Olearia axillaris* and *Grevillea preissii* subsp. *glabrilimba*. Chenille honey myrtle is not suited to long-term pot culture.

Kingsley Dixon

Chenille honey myrtle (*Melaleuca huegelii*).

Grevillea nudiflora

This groundcover has various naturally occurring forms, from totally prostrate to semi-upright. It is most effectively grown as a prostrate plant which can form mats up to 3 metres across. In late winter and spring the red flowers produce an unusual effect at the ends of leafless stems. This species is ideal for growing over high retaining walls and in large gardens, on steep banks and throughout any mixed Australian shrubbery.

David Blumer

Grevillea nudiflora.

Snake bush (*Hemiandra pungens*)

This groundcover can grow up to 1.5 metres in diameter, with the most desirable forms developing into large, dense mats. It has pungent foliage that is sometimes bronze in colour. This species has a range of flower colours from pink to mauve and sometimes white. Snake bush is ideal for growing over high retaining walls and in large gardens, on steep banks and in coastal gardens. A range of colour and foliage forms together creates a great effect *en masse* and they can be planted under an open tree canopy using *Eucalyptus caesia* or *E. pyriformis* or, alternatively, with any coastal planting.

David Blumer

Snake bush (*Hemiandra pungens*).

Scarlet runner (*Kennedia prostrata*)

This large-leafed, open groundcover can grow rapidly up to 1.5 metres in diameter. It has large, bright scarlet pea flowers that provide a high visual impact. A 'must-grow' for coastal gardens, under cultivation this plant can be short-lived and is best replaced every few years. It can be an accent planting or installed among more structural species such as *Chamelaucium uncinatum*, *Banksia burdettii*, *Eremophila glabra*, *Hakea laurina*, *Olearia axillaris* and *Xanthorrhoea preissii*.

David Blumer

Scarlet runner (*Kennedia prostrata*).

Heart-leaf flame pea (*Chorizema cordatum*)

Heart-leaf flame pea is a small shrub to 1 metre high. It is more compact when juvenile. The main flowering is in the early months of spring but spasmodic flowering often occurs before and after this period. The flowers are red, orange and yellow. This species is ideal for planting in areas of semi-shade with species such as *Banksia nivea*, *Adenanthos cunninghamii* and the prostrate form of *Dodonaea ceratocarpa*.

David Blumer

Heart-leaf flame pea (*Chorizema cordatum*).

Scarlet honey myrtle (*Melaleuca fulgens*)

This is a medium shrub growing to 1.5 metres, with narrow, linear leaves. Flowering tends to be a mass show in mid-spring, often on old wood within the shrub. The flowers are large and can be red, pink or orange as well as many shades in between. As this species is open in habit it is a perfect accent plant to mix among a large-scale planting to provide a more open appearance. Where open views and site lines are important, this species can be used to give the garden some structure. It is also suitable as an accent plant in a heathland-type planting, with species such as *Verticordia chrysantha*, *Conostylis candicans*, *Thryptomene denticulata* and *Chorizema cordatum*.

David Blumer

Scarlet honey myrtle (*Melaleuca fulgens*).

Pink rice flower (*Pimelea ferruginea*)

This is an erect, dense shrub which grows to 1 metre high, with dark green foliage. Flowers are predominantly shades of pink and, depending on the form, can appear from August to December. Selecting plants that flower at different times can produce a range of flowering events within your garden landscape. This species is a quick grower and a lover of alkaline soils. It can tolerate coastal influences. It produces an excellent heath-like effect when planted in drifts among plants such as *Scaevola crassifolia*, *Leucophyta brownii* and *Anigozanthos flavidus*.

David Blumer

Pink rice flower (*Pimelea ferruginea*).

Proven complementary species (other than trees, which are described in Chapter 3)		
Annuals (spring)		
Anigozanthos manglesii	*Anigozanthos viridis*	*Lechenaultia biloba*
Rhodanthe chlorocephala subsp. *rosea*	*Rhodanthe manglesii*	*Schoenia filifolia* subsp. *filifolia*
Schoenia filifolia subsp. *subulifolia*	*Trachymene coerulea*	*Xerochrysum bracteatum*
Annuals (summer)		
Crotalaria cunninghamii	*Gomphrena affinis*	*Gomphrena canescens*
Gomphrena cunninghamii	*Ptilotus exaltatus*	*Swainsona formosa*
Climbers		
Hardenbergia comptoniana	*Kennedia lateritia* (formerly *K. macrophylla*)	*Kennedia nigricans*
Groundcovers		
Banksia petiolaris	*Dodonaea ceratocarpa* (prostrate form)	*Hibbertia grossulariifolia*
Medium shrubs		
Acacia aphylla	*Acacia drummondii*	*Acacia glaucoptera*
Acacia merinthophora	*Adenanthos cunninghamii*	*Banksia baueri*
Banksia hookeriana	*Banksia media*	*Beaufortia aestiva*
Beaufortia squarrosa	*Boronia crenulata*	*Eremophila glabra*
Eremophila nivea	*Grevillea bipinnatifida*	*Grevillea petrophiloides*
Grevillea preissii	*Guichenotia macrantha*	*Hakea corymbosa*
Hakea invaginata	*Hakea obtusa*	*Hakea orthorrhyncha*
Hakea platysperma	*Hakea victoriae*	*Olearia axillaris*
Senna artemisioides	*Templetonia retusa*	*Thryptomene denticulata*
Verticordia chrysantha		
Large shrubs		
Acacia denticulosa	*Acacia jibberdingensis*	*Acacia splendens*
Acacia xanthina	*Banksia praemorsa*	*Banksia prionotes*
Banksia sceptrum	*Banksia speciosa*	*Grevillea eriostachya*
Grevillea olivacea	*Hakea bucculenta*	*Hakea scoparia*
Kunzea baxteri	*Kunzea pulchella*	*Melaleuca elliptica*
Melaleuca megacephala	*Melaleuca pentagona*	
Strappy-leafed		
Anigozanthos rufus ('Kings Park Federation Flame')	*Ficinia nodosa*	*Lepidosperma gladiatum*
Macropidia fuliginosa	*Orthrosanthus laxus*	*Patersonia occidentalis*
Structural and iconic		
Macrozamia fraseri	*Santalum acuminatum*	*Xanthorrhoea preissii*

Chapter 3

Small and medium trees

Grady Brand

When you are trying to grow a range of Australian plants within a landscape, trees are often the cause of reduced flowering of any plants that grow within their shadow, as most high impact Australian shrubs and groundcovers require a full sun aspect for best performance. With the exception of the Rottnest teatree, the following selection of trees has been formulated with this in mind. All these trees are proven performers. They have good form and flowers and can be strategically planted within a mixed shrubbery. They are suitable for low maintenance gardens, will attract birds and, unless stated otherwise in the individual descriptions, are suitable for pot culture. The species chosen celebrate the plant diversity found in Western Australia.

Corymbia 'Summer Red'

This eucalypt is a hybrid between two iconic Western Australian species, *Corymbia ficifolia* from the south-west and *C. ptychocarpa*, which grows in the north of the state. Human intervention has brought them together to create a spectacular hybrid which grows up to 5 metres in height and will last for many years. It thrives in all soil types but it is important that the site is free-draining. It produces vibrant red flowers in large heads in summer. A relatively new hybrid to horticulture, 'Summer Red' is well worth trying as a small ornamental single tree, or in a grove planting or a small-scale avenue. Alternatively there are many other selections of *C. ficifolia* which can be planted to produce a similar effect.

David Blumer

Corymbia 'Summer Red'.

Eucalyptus caesia 'Silver Princess'

This graceful, weeping, silver-leafed tree grows to 6 metres high. It has attractive 'minni ritchi' (continuously peeling) bark. There are many different shapes and forms to choose from, to suit any garden setting. Some forms have a more weeping habit with the heights varying. 'Silver Princess' flowers come in a range of shades from pink to red and can be up to 5 centimetres in diameter. It is the signature plant that heralds the winter months. The showy flowers produce nectar, which is an excellent food source for fauna during winter. The buds, stems and leaves are covered in a delicate white powder, which is a strong feature of this species. Plant as a single specimen or group several together to provide screening. *Eucalyptus caesia* has been a garden favourite for many years but is threatened in the wild.

David Blumer

Eucalyptus caesia 'Silver Princess'.

Coolibah (*Eucalyptus victrix*)

This small ornamental tree is iconic due to its mention in the bush ballad *Waltzing Matilda*. It grows to 12 metres high and has a white trunk and glaucous leaves. This species often grows irregularly, giving it character. It lives for 10 years or more. The relatively small, cream flowers appear in summer. The nectar attracts insects and birds and the sweet scent of the flowers will fill a garden. Plant this tree by itself or group several together to provide screening. To produce a distinctive desert garden use coolibah in association with other species that love arid conditions such as *Eucalyptus pachyphylla*, *E. kingsmillii*, *E. youngiana*, *Triodia basedowii*, *T. bynoei*, *Atriplex nummularia*, *Senna artemisioides*, *Hakea rhombales* and *Ptilotus exaltatus*.

David Blumer

Coolibah (*Eucalyptus victrix*).

Pincushion hakea (*Hakea laurina*)

This small ornamental tree is predominantly upright to 6 metres and often multi-branched. Some forms are more dwarfed while others tend to have a weeping habit.

The globular red and cream flowers are spectacular in autumn and winter. This species can be grown as an individual specimen, making a good accent tree within a mixed shrubbery. It is also useful as a small screening plant. It is advisable to plant it in an environment that provides some protection from prevailing winds.

Mack Seale

Pincushion hakea (*Hakea laurina*).

Firewood banksia (*Banksia menziesii*)

This species can take the form of a tree or shrub depending on the seed source, as the species varies genetically throughout the sandplains of Western Australia. The northern form from the Badgingarra region is more commonly the shrub form, ranging in height from 1.3 to 7 metres. The main flowering time is autumn. The terminal flowers may be pink, red, bronze or yellow and may be up to 15 centimetres long by 10 centimetres wide. When in flower this species is a great attractor of nectar-seeking fauna. It is a lover of deep sand but will tolerate heavier soils provided there is good drainage. It can be drift-planted or used as an individual specimen. Its roots are superficial, so care must be given to protect it from any interference. Within large landscapes significant impact can be achieved with a combined planting of this species with *Banksia prionotes*, *B. attenuata*, *Grevillea leucoptera*, *G. candelabroides*, *Hakea bucculenta* and *H. multilineata*.

David Blumer

Firewood banksia (*Banksia menziesii*).

Mulga (*Acacia aneura*)

David Blumer

This grey foliage plant has a range of naturally occurring forms from 1.2 to 10 metres high. Some are shrub-like while others are tree forms that can produce varying landscape outcomes. Originating from the arid regions of Australia, this tree is long-lived, making it a suitable long-term addition to any landscape. Generally the golden-yellow flowers are produced in winter and spring but under cultivation, provided there is reliable water, this species can flower lightly over many months, often doing so spasmodically in response to significant rain events. It has the potential to be used in many arid landscapes. Certain forms are suitable as street trees or for planting in groves or windbreaks. Combine with other desert species such as *Eucalyptus victrix*, *E. youngiana*, *Triodia basedowii*, *T. bynoei* and *Ptilotus exaltatus* to create a display celebrating the flora of these regions.

David Blumer

Mulga (*Acacia aneura*).

Rottnest Island pine (*Callitris preissii*)

This dense, vividly green conifer grows naturally in habitats ranging from limestone to granite and on salt lake margins, displaying a range of growth forms from 1 to 9 metres high and up to 6 metres wide. Produced in late spring the small flowers are yellow-orange, and they are often associated with the phenomenon of pollen clouds being emitted during the warmer spring months. Often grown for its foliage contrast with other more grey-leafed plants, this species makes an ideal windbreak and is a useful plant in coastal areas. These trees grow in varying topographical situations, making them suitable for reclaiming slopes.

David Blumer

Rottnest Island pine (*Callitris preissii*).

Kingsmill's mallee (*Eucalyptus kingsmillii*)

David Blumer

This species thrives in arid conditions and grows in a range of habitats from rocky outcrops to sandy soils. A mallee with varying growth habits from 1.5 to 8 metres high, it also occurs as a compact form, often having pinkish buds. One of the key features of this plant is its ornamental fruits and buds which provide a year-round display. The flowers range in colour from yellow through to pinkish-red and occur in spring. This mallee is a worthy inclusion to any arid landscape. It may be planted in groves or as a specimen for focal points in the garden. Planted among spinifex or with other desert species such as *Eucalyptus youngiana*, *E. pachyphylla* and *Atriplex nummularia*, it creates a celebration of these significantly fruited desert dwellers.

David Blumer

Kingsmill's mallee (*Eucalyptus kingsmillii*).

Bookleaf mallee (*Eucalyptus kruseana*)

David Blumer

This ornamental mallee with grey, orbicular leaves has a very compact habit in its first few years, developing a more open appearance as it ages. Growing up to 3 metres in height, it occurs naturally on granite. Small clusters of yellow flowers in winter and early spring make it a worthy cut flower. Group planting enhances this tree's presence. The species has proven to be very hardy and long-lived under cultivation. Planted on the perimeter of a wooded area, it gives a great understorey appearance and should be considered in any mixed shrub planting.

David Blumer

Bookleaf mallee (*Eucalyptus kruseana*).

Mottlecah (*Eucalyptus macrocarpa*)

In nature this species would be called unruly due to its spreading and sprawling habit. However, under cultivation, regular pruning can tame its growth. The leaves are glaucous and the plant grows to 5 metres high. Often referred to as the largest flowered eucalypt, it bears the most flowers in late spring and early summer, although random flowering occurs throughout the year. The flower colour can vary from pink to shades of red. Effectively used as a foliage plant, group planting will enhance the presence of this species. It has proved to be very hardy and long-lived under cultivation and responds well to formative pruning. Mottlecah is an ideal focal specimen for a predominantly low heath landscape.

David Blumer

Mottlecah (*Eucalyptus macrocarpa*).

Bell-fruited mallee (*Eucalyptus preissiana*)

David Blumer

Open in habit, this small mallee grows to 3 metres high. It is compact during the first years, becoming more open as it grows older. The elliptical leaves are distinctive, as are the bell-shaped fruit which give the common name. This coastal species has large yellow flowers that open during spring, followed by highly ornamental fruit and buds. Effectively used as a foliage plant, it is best planted in groups to enhance its presence. Under cultivation it responds well to formative pruning and then regular pruning to keep it compact. It is ideal for use as cut foliage because of the attractive buds. Plant this mallee with other coast-loving species such as *Scaevola crassifolia*, *Pimelea ferruginea*, *Spinifex longifolia* and *Olearia axillaris*. This species is not suited to long-term pot culture.

David Blumer

Bell-fruited mallee (*Eucalyptus preissiana*).

Pear-fruited mallee (*Eucalyptus pyriformis*)

This small mallee grows to 3 metres in height. It has a very open habit, revealing its flowers boldly at a height where they can easily be admired. From late winter into spring, it produces pendulous flowers on long, down-curved peduncles, with a range of colours but predominantly red or yellow. The flowers are followed by ornamental fruits. Effectively used in a range of planting styles, this tree can be a feature specimen, drift-planted randomly throughout a garden bed or planted as a grove. It responds well to formative pruning and is enhanced when displayed with other large-fruited eucalypts such as *Eucalyptus youngiana* and *E. kingsmillii*.

David Blumer

Pear-fruited mallee (*Eucalyptus pyriformis*).

Coral gum (*Eucalyptus torquata*)

David Blumer

This classic small, shady tree grows to 6 metres high and performs well under cultivation, provided it does not receive excessive summer irrigation. It often performs better on heavier soils or sandy soils over limestone. The canopy is dense and well formed. It flowers over a long period from late winter through to early summer and can have a variety of colours from pink through to red and sometimes cream. The ornamental buds are a strong feature. Planting uses for this species are diverse. It is suitable as a single specimen, a street tree, a small-scale avenue planting or a screen.

David Blumer

Coral gum (*Eucalyptus torquata*).

Grass leaf hakea (*Hakea multilineata*)

This species is a well-shaped, erect, multi-branched tree, often growing 5 to 6 metres in height. The linear leaves are up to 25 centimetres in length. The canopy is dense and the woody fruits, which are retained on the old wood, are a decorative feature in themselves. The flowering spikes are spectacular and grow up to 8 centimetres in length, with shades from pink to red. Where space is limiting this species can be planted as a street tree but it is also suitable for specimen, grove and screen planting. Combined with other species such as *Hakea bucculenta*, *H. laurina*, *H. francisiana* and *H. orthorrhyncha*, it provides a celebration of the more spectacular hakeas which are all large shrubs with showy flowers.

David Blumer

Grass leaf hakea (*Hakea multilineata*).

Salmon white gum (*Eucalyptus lane-poolei*)

This tree grows to 10 metres high. It has an irregularly shaped trunk and its own distinctive form with bark that is mostly smooth, changing from orange-brown to whitish-grey as it ages. The ovoid to globose fruits are also a decorative feature. Cream flowers appear in late summer but are not generally a strong feature. Choose this species if you appreciate the fact that its trunk can grow quite randomly and artistically. It is ideally suited to being planted in a grove with a simple understorey of either *Banksia nivea* or a strappy-leafed plant such as *Baumea preissii*.

David Blumer

Salmon white gum (*Eucalyptus lane-poolei*).

Illyarrie (*Eucalyptus erythrocorys*)

David Blumer

This species has an irregularly shaped trunk. It grows to 8 metres in height, often tending to lean as it matures, heavily laden with large woody fruits. Large, spectacular, yellow flowers adorned with a red operculum produce a vivid effect in late summer. This tree deserves an accolade as the signature species for the Swan Coastal Plain in Perth, Western Australia, as it is ideally suited to coastal conditions and starts to flower in the middle of summer when little else is in bloom. Illyarrie is most effectively planted in groups or large groves, with an understorey which includes plants such as *Conostylis candicans* and *Grevillea thelemanniana* subsp. *glabrilimba*. This species is not suited to long-term pot culture.

David Blumer

Illyarrie (*Eucalyptus erythrocorys*).

Wirewood (*Acacia coriacea*)

This species has a variety of forms ranging from compact, large shrubs (which tend to prefer coastal environments) to erect, shapely trees growing to 10 metres in height. The foliage is grey-green and linear. Yellow globular flowers appear in winter, followed by long ornamental seedpods. Wirewood is one of the longer-lived wattles and is valued for its distinctive lime-green foliage which provides great contrast and focal points within a landscape. It is ideally used as a street tree, a screen, in a grove or as an individual specimen within a landscape with an arid flora theme.

David Blumer

Wirewood (*Acacia coriacea*).

Darling Range ghost gum (*Eucalyptus laeliae*)

This is a well shaped tree growing up to 20 metres on heavier soils, although it tends to grow to slightly less (12 metres) in free-draining sandy soils. The bark is smooth and powdery, usually white but changing to a butter yellow in autumn. Small and relatively insignificant flowers appear in summer. Its iconic white trunk is a feature that engenders interest and the tree makes a great addition to any landscape as a specimen, grove or small-scale avenue planting.

David Blumer

Darling Range ghost gum (*Eucalyptus laeliae*).

Bushy yate (*Eucalyptus lehmannii*)

Often multi-trunked, this small well shaped mallee grows to 5 metres in height and has smooth, light pink to grey bark, often with accumulated older bark at the trunk base. Highly decorative fruits are a feature. For many years this species was confused with *Eucalyptus conferruminata* but *E. lehmannii* is now considered more versatile as its flowers and fruits are more ornate. Its flowers are fused into a tight ball, which produces a spectacular effect in late summer months. The inflorescences are green and borne from highly ornamental, horn-like clusters. This species is a great addition to any landscape as the ornamental shape and highly decorative flowers, buds and fruits will add year-round attraction. It could be planted as a small screen, as an individual specimen or in a grove.

Luke Sweedman

Bushy yate (*Eucalyptus lehmannii*).

Rottnest teatree (*Melaleuca lanceolata*)

David Blumer

This species is diverse in habit, but is most commonly sold as a tree form growing up to 8 metres tall. Unlike all the other species described in this chapter, Rottnest teatree does create dense shade under which very little else will grow. The flowers are cream, often occurring over a long period but tending to be spasmodic. This species is a classic lover of coastal regions and is ideal for windbreaks or enhancing large open spaces. Planting in large groves produces an enchanting forest effect within a parkland environment. This species is not suited to long-term pot culture.

David Blumer

Rottnest teatree (*Melaleuca lanceolata*).

Other medium trees to consider		
Corymbia calophylla	*Eucalyptus burracoppinensis*	*Eucalyptus conferruminata*
Eucalyptus crucis subsp. *crucis*	*Eucalyptus densa* subsp. *densa*	*Eucalyptus desmondensis*
Eucalyptus drummondii	*Eucalyptus formanii*	*Eucalyptus forrestiana*
Eucalyptus macrandra	*Eucalyptus marginata*	*Eucalyptus megacarpa*
Eucalyptus megacornuta	*Eucalyptus nutans*	*Eucalyptus orbifolia*
Eucalyptus platypus	*Eucalyptus pleurocarpa*	*Eucalyptus roycei*
Eucalyptus sargentii	*Eucalyptus sepulcralis*	*Eucalyptus spathulata*
Eucalyptus tetraptera	*Eucalyptus todtiana*	*Eucalyptus wandoo*
Eucalyptus websteriana	*Eucalyptus woodwardii*	*Melaleuca cuticularis*
Melaleuca globifera	*Melaleuca leucadendra*	*Melaleuca nesophila*
Nuytsia floribunda	*Xylomelum angustifolium*	

Chapter 4

Arboriculture

Jeremy Thomas

Trees are significant and dynamic features of a landscape and require specific consideration and management. Whether growing your own trees, buying them from a nursery or having them contract-grown, it is critical to understand how the mature tree will fit into the landscape and to choose only quality planting material.

Careful selection of species matters. The right tree in the right place is a valuable long-term asset and provides many benefits including the provision of shade and a place for birds to roost or nest. Understanding the natural growth characteristics of a tree is also important in order to assess whether there will be enough space above and below ground for it to reach its full size in the chosen location. Familiarise yourself with the site and environmental conditions (soils, water, temperature and sun or shade) that the species prefers for healthy development.

Consulting a professional arborist may be necessary to obtain advice about any natural structural weakness or pest and disease issues that may be associated with the selected species. Discuss your requirements with the arborist and determine the level of care expected during the tree's life in terms of formative pruning, ongoing remedial maintenance and water and fertilising requirements.

Before choosing the species and form of tree that best suits the desired situation – especially in public landscapes – consider the intent or design of the planting position, the long-term cost of maintenance and the equipment required to achieve those objectives.

Buying trees

Growing or selecting good quality stock will help prevent common disorders later in life. When choosing trees, you should consider the following guidelines:

- Ensure that the rootstock is free from defects such as excessive root pruning or girdling.
- Avoid stems with clustered branching from incorrect pruning, as well as included bark (non-connective tissue between co-dominant stems) within stem unions and dual leading stems as these may split and fail in time.
- Choose material that indicates good health and vigour, is free from pests and has no sign of disease.
- Avoid trees left in nursery containers for too long. These eventually develop root girdling defects, a leading cause of tree decline and premature removal. Large trees grown in nursery tubes, pots and bags often require careful root manipulation during planting to free the roots from their container. Although the cutting of large roots is possible when freeing them from a container, tolerance to this is often species dependent and, unless previous experience indicates otherwise, it is usually best to use another plant to avoid long-term problems.

Jeremy Thomas

A 110-year-old Australian red cedar (*Toona ciliata*) from eastern Australia growing in Kings Park and Botanic Garden is 10 metres tall and 15 metres wide.

Growing your own trees

At Kings Park all the trees destined for parkland planting are grown in the nursery within the grounds. Growing your own trees can be a rewarding experience and it is not as difficult as some people believe, provided the following simple guidelines are followed:

- Buy quality seed from a reliable source.
- Be aware that the number of seeds to sow will depend on seed viability and germinability.
- Thinning and selection of emerging seedlings requires the careful cutting of inferior seedlings below their cotyledons.
- Pricking out and/or handling the seedling roughly may cause damage to the root radical and give rise to future establishment issues.
- It is preferable to direct sow into the largest container possible and one that allows the new tree to grow without the need to re-pot.
- Use a potting mix with slow release fertiliser included that best meets the tree's requirements. Incorporate this mix with your existing soil when eventually planting the tree in the landscape to offer the tree good soil organics during early development.
- Provide the tree with a reliable and regular water supply. Insufficient water or sudden changes to water availability, especially in warmer weather, may contribute towards reduced health and even plant death.

As a general rule, sowing the seed of southern Australian species in winter and growing the plant in its pot for 10 to 12 months before planting out the following winter produces the best results.

Jeremy Thomas

The tree nursery at Kings Park.

Nursery care

In standard pots and bags, roots will develop laterally and then encircle the pot. If these roots are left to mature they will continue to grow in this way even after they are in the ground. Once planted, some roots may escape this circular formation but, while some trees may appear to be growing well, more may experience early poor vigour, subsequent health issues and destabilisation and/or premature death.

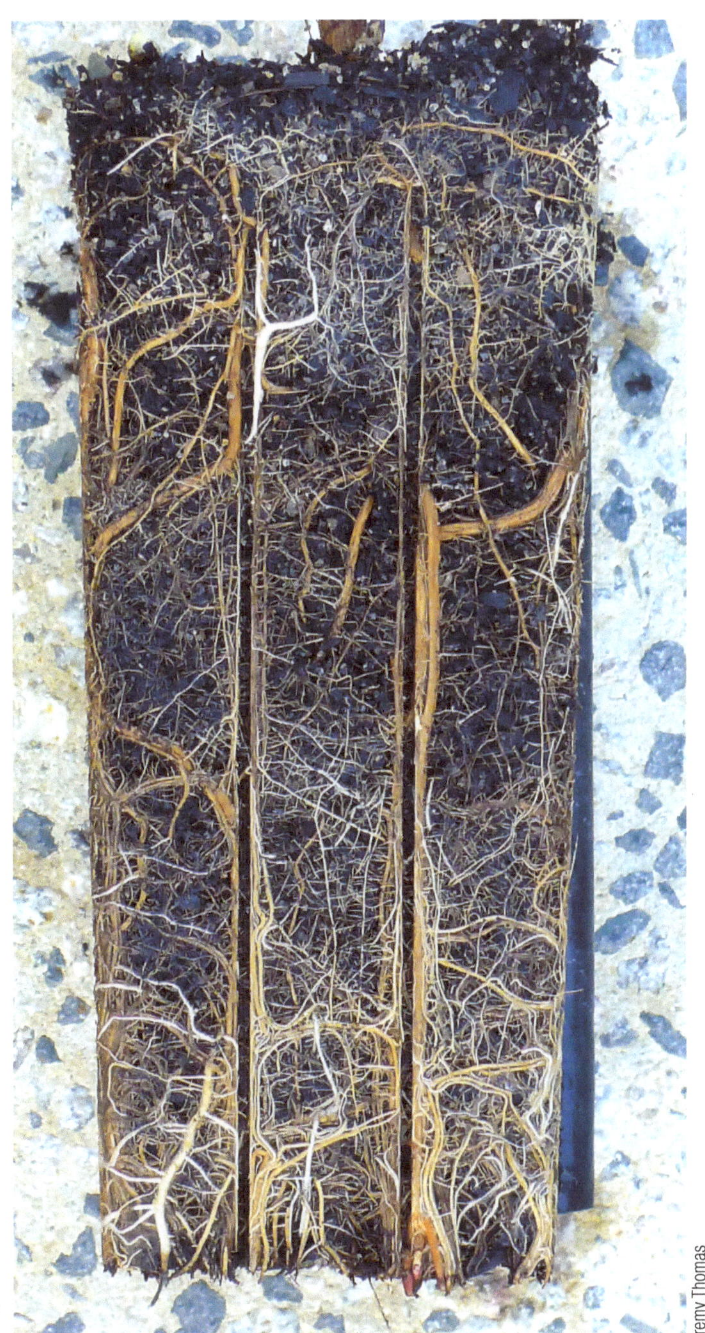
Jeremy Thomas

An early root defect in a forestry tube.

Advances in nursery pot development have greatly helped to reduce the incidence of root-binding and girdling. Using three-dimensional air root-pruning (3DARP) pots with conical perforations along the walls of the pot encourages vigorous root growth. As the root tip protrudes through the conical opening it is exposed to light and air, which cause the root tip to dehydrate. Secondary root fibre arises from behind the dehydrated root tip and this style of growth occurs throughout the root ball, offering an increase in healthy root mass.

When using 3DARP pots, it is better to sow directly into the pot that the plant will grow in, and thin out by cutting surplus seedlings below the cotyledon leaves to ensure there is no disturbance to the roots of the remaining seedlings. If this can't be done, then sowing into a smaller 3DARP pot and carefully potting into a larger pot as necessary also gives good long-term results.

Jeremy Thomas

Three-dimensional air root-pruning pot.

Corrective pruning should commence at the nursery stage. Identify potential defects such as co-dominant leading stems, included bark within stem unions and clustered branching where conflict of space will occur. However, when pruning retain as much photosynthetic material as possible and retain all lower main stem growth as this is important for main stem development and strength. In most cases after planting out, this lower stem growth should remain for at least the first season before being removed.

All pruning methods should be consistent with Australian Standards 'Pruning of Amenity Trees AS 4373–2007'.

Tree planting

Preparing the site before planting will ensure the best start to a tree's life, so familiarise yourself with the planting site in order to better understand the pH, nutrient levels, soluble salt content, past use and any known diseases.

Excavate the planting hole up to two to three times the width of the tree root ball. The sides should slope outwards from the bottom of the excavation at 25 degrees, particularly in heavier soils. The depth should equal that of the root ball. Planting trees too high will cause some root decline because the roots are exposed to air and sunlight, while planting too deep may cause root damage and incite main stem collar rot in time.

Orientate the tree to present its most desirable face to the place from which it will mainly be viewed. Where bent main stems exist, ensure the head of the tree is straight, as opposed to having a straight main stem. This way, as the tree grows towards the natural sunlight no further bending moments will be produced along the main stem.

Backfill the hole with a mixture of existing and new soil, watering incrementally to avoid air pockets where fungi may proliferate. Firm the soil as you progress. Create a 7 to 8 centimetre high basin around the tree to encourage water retention around the root ball. In heavier soils it may be necessary to reduce water volumes to avoid soil saturation that may lead to the tree shifting.

Always apply composted wood mulch over the surface soil to a thickness of about 5 centimetres and ensure the main stem is not in contact with the mulch. Avoid mulches that have the propensity to form a dry crust, such as those containing large amounts of finely shredded bark and peat products, which prevents water penetration.

Emily Livingstone

An eight-month-old marri (*Corymbia calophylla*) in a three-dimensional air root-pruning pot.

Staking and guying

Incorrect or superfluous staking and guying may cause structural weakness through over-support. Stake and guy trees only when absolutely necessary – for example, to protect against vandalism or to provide stability in windy sites. Stakes and guy ropes should be removed as soon as possible to encourage the tree to develop self-support.

Since 2006 thousands of trees from 3DARP pots have been planted in Kings Park. The quality of these trees and their self-supporting nature has meant none has required staking. This represents a significant saving in labour costs and has the added advantage of reducing the effect of damage that is often seen through incorrect staking.

In-ground care

Once a tree is planted it is important to monitor progress during the first season. During this time the tree may experience some subtle changes to condition while adjusting to its new environment.

Dead and diseased branches can be removed at any time of the year. Live branches of evergreen trees are best retained until after a flush of growth and/or flowering, while branches of deciduous trees are best pruned during their dormancy period. Do not reduce the height of a new tree or prune indiscriminately to remove or reduce stems. Annual maintenance should never involve significantly pruning the tree, except for removing dead branches. The following guidelines may be useful:

- Year 1 (in the nursery): Apply corrective pruning to remove defects and promote integrity. Depending on species the tree may be 1 to 1.5 metres tall at the end of this stage.
- Year 2: As this is the first season in-ground, minimal pruning is required. Minor directional pruning will promote good structural form within the allocated growing space. Remove senescing basal growth of the main stem, remove dead wood and replenish mulch.
- Year 3: Minor pruning is required such as canopy uplifting for clearance, removing dead wood and removing any excessive epicormic regrowth. Replenish mulch. Depending on species the tree may be 4 to 5 metres tall at this stage.

- Year 5: By now the tree is likely to be of a size that requires professional attention. In general, every tree should be inspected annually or biennially. Servicing the tree every 3 to 5 years thereafter may be required to maintain health and form.

Pruning best practice

Target pruning is an industry standard for correct stem removal. Target pruning recognises the branch bark ridge and branch collar. The branch collar contains main stem tissues and branch tissues and it is within this region that wound wood (callus) is produced. Damage to the branch collar (flush cutting) can instigate decay and cracks, further compromising health and structural integrity. Do not apply wound sealants as they do not prevent decay and, in some cases, may be detrimental to natural, healthy wound wood development.

Through semi-maturity and into maturity it will be prudent to engage a qualified arborist or tree surgeon to maintain a tree. Tree surgery often requires a specialised approach and may call for a variety of specific actions in order to maintain tree health, structural integrity and overarching risk matters.

Root investigation and protection

Kings Park uses a specialised air tool to excavate and investigate without the need to cut and destroy important tree roots. The AIR-SPADE® is a specialised hand-held tool that produces a stream of supersonic air moving at Mach 2. The AIR-SPADE® effectively penetrates and dislodges most types of soil, but is harmless to non-porous objects such as plant roots and buried pipes or cables.

Jeremy Thomas

Using AIR-SPADE® to install irrigation in Kings Park.

Jeremy Thomas

Retaining roots while enabling the installation of services.

Tree preservation through relocation

Kings Park has been transplanting significant trees for over 20 years with assistance from local arboriculture specialists. Transplanting allows the park to save many trees under threat from construction and development and to preserve them for future generations to enjoy.

Transplanting mature trees is usually expensive and time-consuming. However, when the intrinsic value of the tree and the benefit it provides are considered, the loss of amenity resulting from removal of the tree often makes tree relocation cost-effective and worthwhile. Relocation is a specialised operation that can take up to 18 months of preparation time and requires knowledge of the tree's ability to tolerate trauma and grow in a new site.

Jeremy Thomas

Relocated red-flowering gum (*Corymbia ficifolia*) in Kings Park.

The relocation process involves determining which roots within the root ball will be severed, which will be only partially severed, and which need to be fully retained to ensure ongoing tree stability. Before lifting, the prepared root ball is protected with a geo-textile material and packed with a special blend of materials to promote and protect new root growth.

Trees which in the past were considered very difficult or even impossible to relocate without a high risk of mortality (including many species in the Myrtaceae), are now regularly relocated in Kings Park. Successful relocations include an 85-year-old red-flowering gum (*Corymbia ficifolia*) and a 15-year-old coolibah (*Eucalyptus victrix*).

Jeremy Thomas

Relocated coolibah (*Eucalyptus victrix*) in Kings Park.

The relocation of a 750-year-old boab (*Adansonia gregorii*) in 2008 was done under less than ideal conditions, with only 3 months' notice about the desire of local Aboriginal people to relocate the tree to Kings Park in advance of a bridge construction project in the Kimberley region in northern Western Australia.

Preparation of the boab *in situ* included selective pruning and root management in readiness for its 3500 kilometre journey. Preparation in Kings Park included removing 70 cubic metres of existing soil and replacing it with coarse river sand similar to that found in the Kimberley region. Other works included installation of a customised summer watering system and protection of the root zone from winter rains, to allow management of the soil environment to replicate Kimberley conditions as closely as possible.

Jeremy Thomas

Removing the 750-year-old boab tree (*Adansonia gregorii*) in readiness for its 3500 kilometre journey to Kings Park.

Jeremy Thomas

The successfully relocated 750-year old boab in Kings Park.

Relocation of mature trees is not always feasible or economic but experience in Kings Park shows that a successful transplant can provide long-term benefits to the landscape by prolonging individual tree life. It also provides the opportunity to explain the techniques used and the reason for the relocation to park visitors.

Chapter 5

Propagation

Amanda Shade and Mark Webb

The two main nursery techniques used to propagate Australian native plants are seed germination and production from cuttings. Seed-grown material produces plants with the genetic make-up of both parents so that every plant is slightly different genetically from its sibling, while cuttings produce plants that are clonal, that is, they have the same characteristics as the plant from which the cuttings were taken.

Other less common propagation techniques include division, grafting and tissue culture. Plant division is used to propagate genera that have rhizomes, stolons or bulbs such as *Patersonia*, *Conostylis*, *Orthrosanthus* and *Anigozanthos*. Grafting is widely used for other ornamentals such as roses and especially for fruit crops, and is becoming more commonly used on native plants. Tissue culture (see Chapter 7) is an expensive but successful laboratory-based propagation technique that is used when more traditional techniques have not been successful, or when rapid multiplication of clonal material is required.

Soil mixes for seed germination and cuttings

Any clean, well drained medium incorporating sand and organic matter at suitable ratios can be used for sowing *seed*. A mix successfully used for many years at Kings Park and Botanic Garden is a combination of composted jarrah sawdust (or another suitable composted organic material), with river sand and fine gravel in a ratio of 2:1:0.5, with slow release fertiliser added at label rates and pH adjusted to a level of 5.8 to 6.8.

Any mix used for *cuttings* must have a good balance between drainage and water-holding capacity. Clean, coarse river sand, peat moss and perlite in a ratio of 2:1:1 with no added fertiliser is a reliable mix regularly used at Kings Park for native plants.

David Blumer

Composting jarrah sawdust.

David Blumer

Kings Park standard mix for cuttings.

Seed propagation

Top tips for best results:

- Keep good records including seed age, sowing dates and any pre-treatments used.
- Use a high quality sowing mix.
- Maintain good hygiene.
- Don't cover the seed to a depth more than twice the diameter of the seed.
- Don't overwater.

Seed should only be obtained from a supplier who can guarantee that the seed has been legally and sustainably collected from healthy plants, and cleaned and stored under optimum conditions.

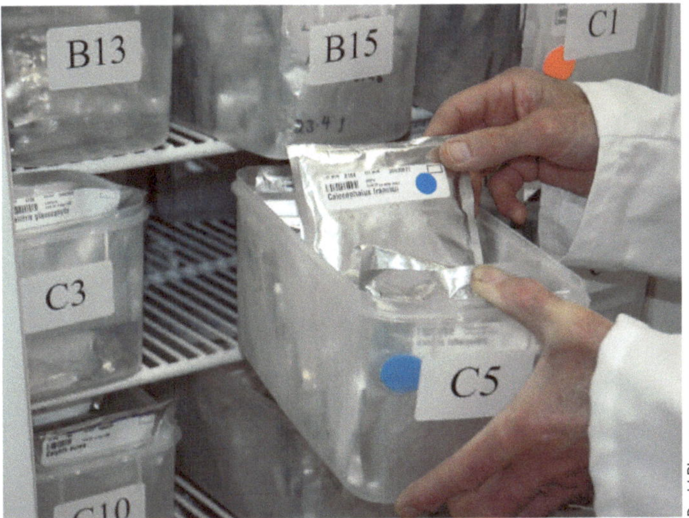

David Blumer

Properly labelled seed packets being removed for seed sowing.

Australian plant species can be grouped into those that store mature seed in the plant canopy (serotinous or bradysporous) and release them in response to a disturbance event (usually fire), or those that release ripe seed each year into the soil seed bank (geosporous). Generally, seeds of serotinous species are non-dormant and germinate readily. In contrast, seeds of many geosporous species are dormant when mature and some pre-treatment is usually required for successful germination.

Currently, about one-third of Western Australian native plant species are difficult to propagate from seed for a variety of reasons including low seed set, poor seed germination and seed dormancy mechanisms. Determining the right conditions for seed germination requires consideration of several factors including seed maturity, dormancy type (for example, physical or physiological), seed storage conditions and storage period.

Various seed treatments prior to sowing can improve germination outcomes in many species. For many serotinous genera, including *Banksia* and *Eucalyptus,* no seed treatment is usually required to achieve reasonable germination. In contrast, for mainly geosporous genera, including *Acacia*, *Anigozanthos*, *Grevillea* and *Swainsona*, seed pre-treatment is usually necessary.

David Blumer

Seed of different species ready for treatment.

The most common seed pre-treatments used at Kings Park and Botanic Garden for Australian native plants are:

- **Hot water treatment:** Many genera within the Fabaceae, such as *Acacia*, *Senna*, *Swainsona* and *Bossiaea*, have a seed coat that is impermeable to water (physical dormancy). Soaking these seeds in near-boiling water for between 30 seconds and 5 minutes, depending on the species, will allow the seed to take in water and germinate readily. Species of *Acacia* and most physically dormant species in *Malvaceae*, *Rhamnaceae* and *Sapindaceae* with harder seed coats germinate well when placed in hot water for 2 to 5 minutes. For other species with softer seed coats, such as those of *Bossiaea*, *Daviesia* and *Gompholobium*, placing them in hot water for only 30 seconds to 1 minute is recommended. Alternatively, removing or 'nicking' a small section of the seed coat using a scalpel, file or a pair of scissors for small quantities of seed (or a mechanised scarifier for larger quantities) and soaking the seed in room-temperature water overnight before sowing will achieve a similar result.

David Blumer

Removing or 'nicking' section of seed coat of *Cochlospermum fraseri*.

- **Smoke treatment:** Soak seed of responsive species in a 10 per cent smoke water solution at room temperature for approximately 24 hours before sowing. Alternatively, untreated seed can be well watered in the seedling tray with a 10 per cent smoke water solution and then left in a cool position for 24 hours before continuing with normal watering. Vermiculite inoculated with smoke and applied as a light covering over the seed and then watered is particularly useful in germinating very small seed that is often difficult to sow after being soaked in a smoke water solution.

David Blumer

Ten per cent smoke water solution being added to seed of *Patersonia occidentalis*.

- **Gibberellic acid (GA):** This plant growth regulator is useful for seed with physiological (embryo-based) dormancy. Soak seed of responsive species in a 1 per cent GA solution for approximately 24 hours before sowing.

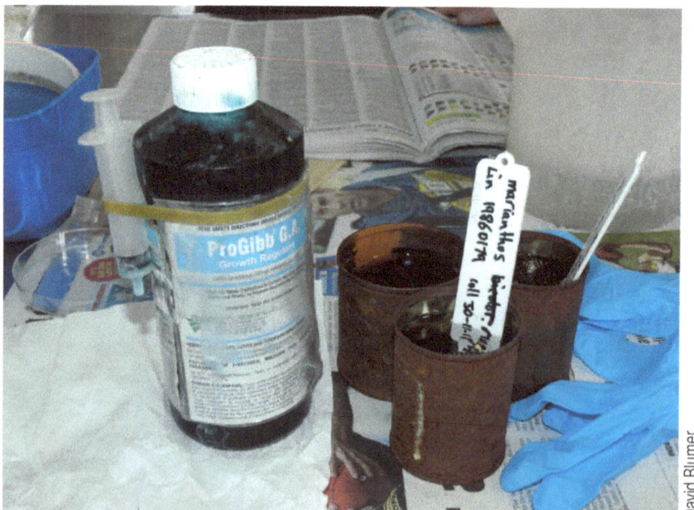

David Blumer

Seed of *Marianthus ringens* before being soaked in 1 per cent Gibberellic acid (GA) solution for 24 hours.

For some species, germination can be further enhanced by combining seed pre-treatments. For example, the germination of fresh seed of *Anigozanthos manglesii* and other kangaroo paws can be significantly improved by placing seed in an oven (to simulate seed ageing) at 100°C for 3 hours then smoke-treating it before sowing. Seed of kangaroo paws stored at room temperature and older than 12 months of age does not require any additional heat treatment. Seed from genera that have complex dormancy-breaking requirements, such as *Hibbertia*, can give better results after exposure to both smoke and GA treatments.

Many Western Australian natives have very specific temperature requirements for dormancy break and germination. Most species naturally germinate during a period of reliable rainfall but many can also be germinated out of season with the right nursery conditions and appropriate cold or heat treatments. For mainly southern Australian genera such as *Banksia,* which are being germinated outside their normal seasonal pattern, sown trays are placed in a cool room at 15 to 18°C for 25 to 35 days or until germination has been achieved, before placing them in a glasshouse. For northern heat-loving species such as *Swainsona*, sown trays are placed in a glasshouse with an ambient temperature of 28°C and a bottom heat of 30°C.

David Blumer

Emerging seedlings of banksia.

To produce seedlings in Western Australia for an early winter transplant, seed is usually sown in late spring or early summer. Fine seed such as from *Beaufortia*, *Kunzea* or *Melaleuca* can be spread out over the soil surface on a tray and then lightly covered with vermiculite. Larger seed such as that from some *Acacia*, *Hakea* or *Grevillea* is usually planted at a depth of about one-and-a-half to two times the width of the seed and then lightly covered with vermiculite. Vermiculite is a light, sterile material that not only allows air and water penetration but can also provide a physical barrier to glasshouse pests such as fungal gnats. For 'winged' seed such as *Banksia,* the seed is usually planted flat or with the point down. After sowing, place the tray in a propagation house and use fine misting sprays that keep the soil surface moist but not wet. Hand-water very lightly if necessary.

David Blumer

Vermiculite covered seed being lightly watered.

Seed will usually germinate 2 to 4 weeks after sowing but can take much longer depending on species and conditions. Sometimes seed is sown too thickly or germinates better than expected. In these cases, thinning out soon after germination prevents the seedlings from becoming too spindly and makes later pricking out easier. Carefully prick out the seedlings when they have two true leaves or are large enough to place easily into a 5 centimetre diameter tube or a growing-on pot containing a moist but well drained mix. It is critical to ensure that the root is carefully placed into a tube which is longer than the root, then carefully press the soil around the seedling and water in well. If the seedling root is badly bent during pricking out, it can result in a 'J' or kinked root that can lead to problems later on in the plant's life. For some fine-rooted plants prone to 'J' roots, the seedling should be laid down the length of a tube and the mix carefully poured in to fill up the tube. Prick out into containers that allow for air-pruning. Air-pruning occurs as the roots grow out of the pot and stop when they are exposed directly to air. Root-bound plants are a result of keeping seedlings for extended periods in too small a pot or without adequate air-pruning.

David Blumer

Carefully pour soil mix around seedling to encourage good root growth.

Cuttings

Top tips for best results:

- Keep good records including type of material, planting dates and hormone rate.
- Use healthy material and maintain good hygiene.
- Use a good quality potting mix.
- Don't let cutting material dry out before use.
- Smaller cuttings are usually better than larger cuttings.

Vegetative propagation is usually done when seed germination is unreliable or insufficient seed is available, or when clonal material is required. Vegetative propagation is mainly done from stem cuttings, but leaf or root cuttings can also be used. For stem cuttings, the tip ends of semi-hardwood growth (the hardened young wood present after a growth flush, especially after flowering) usually give good results. For many species, spring or early summer is the best time to take cuttings. With experience, cuttings can be taken at any time of the year provided suitable growing-on conditions are available, although strike rates might be lower and growth rates slower than during the preferred times. Cuttings should be taken from healthy plants in a cooler part of the day, labelled, placed into 'breathable' plastic bags and stored in a cooler bag, or placed into water and stored in a cool place and regularly sprayed with a fine mist to prevent them drying out before preparation in hygienic, cool conditions. If the cuttings are not to be used immediately they can be wrapped in moist newspaper, placed inside a breathable plastic bag and stored in a refrigerator for several days.

Cuttings for most species should be 5 to 10 centimetres in length (this will depend on factors such as the habit and internodal length) and/or have at least two nodes above and below the intended planting level. Any very soft tip growth and leaves below the anticipated soil level at planting should be carefully removed and very large leaves trimmed. Proper record keeping will help to identify good mother plants, suitable propagation techniques, and the best times for propagating different species and varieties.

A standard mix suitable for most species contains two parts of river sand and one part each of peat moss and perlite. For *Hemigenia* or *Olearia*, a mix with more organic matter such as composted jarrah sawdust, combined with perlite in a ratio of 1:1 and low phosphorus slow release fertiliser, gives good results. Rockwool has proven very successful for striking species from dry inland areas of Western Australia, especially those with soft or furry foliage such as *Dicrastylis*, *Eremophila* and *Lachnostachys*. Using the standard mix in individual fibre pots can improve results for striking *Grevillea*, *Hakea*, *Lambertia* and *Conospermum* (see Table 1 for comparisons with *Grevillea*).

Depending on species, type of material and time of year, different treatments can be applied to cuttings to improve the strike rate. For most semi-hardwood cuttings, the use of an auxin (a plant hormone) such as indole-3-butyric acid (IBA) at 3000 ppm as a gel, liquid or powder will improve the strike rate. For information on the average days to strike a range of species, see Table 5 on page 54.

Hardwood cuttings are not generally used for most Australian plants, but can sometimes be suitable for some Proteaceae such as *Banksia* and *Grevillea*. For hardwood cuttings, hormone rates at 8000 ppm IBA equivalent should be used.

David Blumer

Cuttings of *Dicrastylis incana* (left) and *Quoya verbascina* (right) in rockwool cubes.

David Blumer

Dipping the end of a cutting in 3000 ppm rooting hormone.

Cuttings of *Conospermum stoechadis* in fibre pots.

Table 1: Effect of different media on selected Western Australian species

Species	Medium	Average strike rate	Highest strike rate
Dicrastylis fulva	Sand:perlite:peat moss (2:1:1) in punnets	35%	91%
Dicrastylis fulva	Rockwool cubes	58%	100%
Lachnostachys verbascifolia	Sand:perlite:peat moss (2:1:1) in punnets	15%	36%
Lachnostachys verbascifolia	Rockwool cubes	41%	88%
Grevillea flexuosa	Sand:perlite:peat moss (2:1:1) in punnets	23%	47%
Grevillea flexuosa	Sand:perlite:peat moss (2:1:1) in individual fibre pots	61%	94%
Hemigenia ramosissima	Sand:perlite:peat moss (2:1:1) in punnets	0%	0%
Hemigenia ramosissima	Sand:peat moss:jarrah sawdust (1:1:1) plus fertiliser	80%	100%

When ready to strike the cuttings, make a suitable sized hole with a clean stick in the slightly moist potting mix, dip the base of the cutting in a hormone gel, liquid or powder and place it in the hole so that half to one-third of the cutting is below the surface. Carefully press around the cutting to make sure it has good contact with the soil and air pockets are minimised. Maintain adequate space between each cutting to prevent overcrowding and continue this process until the tray is full. Regularly moisten the newly planted cuttings with water mist from a spray bottle. Place the tray in a propagation area with an air temperature of 22 to 24°C and a bottom heat of 24 to 26°C. Best results are achieved with misters that come on regularly to keep humidity high and prevent the cuttings from drying out but without making them too wet, which can lead to rotting.

Leaf cuttings can be used for species with thick, fleshy leaves, such as some *Goodenia*, or other species without obvious stems, such as some of the basal rosette forms of *Stylidium*. Leaf cuttings are done the same way as for stem cuttings except that the whole leaf including the petiole (if present) is used.

When the majority of cuttings have struck (as usually evidenced by roots being seen at the air-holes at the base of the punnet or tray), they can be carefully potted into larger pots and then transferred to a glasshouse or shade-cloth-covered hardening off area, before being placed in full sun until ready for use. Depending on ambient temperatures, plants might take as little as 3 weeks or as long as 2 months to harden off before they can be moved outside into full sun. In very hot climates, plants may need to be kept under 30 per cent shade-cloth during the hotter months before planting out in the cooler season.

David Blumer

Regular misting of cuttings prevents them drying out.

Grafting

Grafting can be done when:

- Seed or cutting propagation has not been successful.
- Very limited vegetative material is available.
- Characteristics (especially flower colour) of the clonal selection need to be maintained.
- The plant does not have a strong or vigorous root system.
- The plant does not grow well in the soil at the intended planting site or is susceptible to pathogens such as nematodes or phytophthora.

The soil at Kings Park's State Botanic Garden is sandy and slightly alkaline, so displaying species from other parts of Western Australia that need different soil conditions for optimum results can be challenging. Sometimes entire genera that normally grow on more acid soils, such as *Verticordia,* provide better display outcomes on alkaline soils when grafted. However, they may still require soil amelioration or foliar trace element applications to avoid obvious nutrient deficiencies, especially iron deficiency.

Grafting involves attaching a section of stem from one plant (scion) onto the stem of another plant (rootstock) so that the scion becomes part of the rootstock. Consequently, these grafted plants can be grown on a range of soil types or where there is a known infestation of certain pests such as nematodes, or diseases such as phytophthora. Normally, the more closely related the scion and rootstock species are to each other, the more likely the graft will be successful. However, this is not always the case and genetically distant species such as *Dicrastylis fulva* (scion) and *Westringia dampieri* (rootstock) can sometimes result in a successful graft.

Grafting is usually done by hand onto established rootstocks. Wedge graft and splice or whip graft are the most common grafting techniques used for Australian plants. A 'cutting graft', where propagation of the rootstock and grafting of the scion is done at the same time, can be done when the rootstock strikes quickly and reliably. Generally, grafts of Australian natives are done on young plants with stem diameters up to 4 millimetres. The grafting technique selected needs to be easy to perform and give a strong and stable graft. Grafting can give exciting outcomes but it can also be a frustrating experience and sometimes, despite all other factors appearing to be equal, poor results can be obtained using techniques that would normally give a good outcome. For best results, plants just grafted are placed inside a propagation tent at 20 to 28°C, with a fogging unit that maintains at least 90 per cent humidity. Capillary matting over a heat mat set at 24 to 26°C will avoid the need for overhead watering and ensure the grafts do not become too wet.

The most critical factor in successful grafting between compatible plants is to have good contact between the cambium tissue of the scion and the rootstock. This usually means that the diameters of the scion and the rootstock must be similar. When the scion stem diameter is smaller than the rootstock stem diameter, a wedge graft is the preferred technique.

Genera grafted at Kings Park include *Eremophila* (Scrophulariaceae), *Verticordia* and *Darwinia* (Myrtaceae), *Boronia* (Rutaceae), *Pimelea* (Thymelaeaceae), *Prostanthera* and *Hemiandra* (Lamiaceae) and *Grevillea* (Proteaceae). Species regularly grafted include *Corymbia ficifolia, Swainsona formosa, Microcorys eremophiloides, Pityrodia scabra, Diplolaena angustifolia* and *Dicrastylis fulva.*

Rootstock selection is not limited to within the same genera or even the same continent and thus can provide a wide range of potential rootstocks for different situations (see Table 2). Preliminary trials with *Solanum orbiculatum*, which is native to Western Australia, onto *Solanum rantonnetii* 'Royal Robe' from South America, have shown excellent compatibility with encouraging long-term results.

Table 2: Successful scion–rootstock combinations of selected Western Australian species

Family	Scion species	Rootstock species	Rootstock qualities
Scrophulariaceae	*Eremophila nivea*	*Myoporum montanum*	Hardy; adaptable to a range of conditions
Lamiaceae	*Prostanthera magnifica*	*Westringia dampieri*	Hardy; adaptable to a range of conditions
Thymelaeaceae	*Pimelea physodes*	*Pimelea ferruginea*	Excellent for coastal conditions
Myrtaceae	*Darwinia meeboldii*	*Darwinia citriodora*	Hardy; good in coastal soils
Myrtaceae	*Verticordia helichrysantha*	*Chamelaucium uncinatum* hybrids	Some forms show tolerance to phytophthora
Rutaceae	*Boronia megastigma*	*Boronia clavata*	Hardy, adaptable to a range of conditions
Solanaceae	*Solanum orbiculatum*	*Solanum rantonnetii*	Exotic species; easy to propagate
Rutaceae	*Diplolaena angustifolia*	*Correa pulchella*	Eastern Australian species; easy to propagate

Wedge graft – step by step

Required tools for grafting include sharp cutting instruments (such as finely pointed secateurs, scalpels or budding knives), grafting tape or laboratory film for wrapping grafts, methylated spirits or bleach for sterilising tools, and a plastic hood or cover to provide a humid environment if suitable glasshouse facilities are not available (see next page).

For a wedge graft, match up the scion material with stems of similar diameter from the rootstock so maximum cambium tissue contact is achieved.

The scion is usually prepared first. Trim the material to about the same length as if preparing it for a cutting. Remove very soft tip growth or excess side shoots, and trim any larger leaves. Using a sharp, sterilised blade, slice one side of a 'V' shape at the bottom of the scion section. Turn the scion around and repeat to match the other side (see diagram 1 on next page). Ensure the cuts are straight and clean to maximise contact with the rootstock tissue.

The length of the 'V' or wedge will depend on the species, but longer is usually better than shorter, as it enables more tissue to make contact, and allows for better stability once the graft has taken.

To prepare the rootstock, remove a stem of the plant to a desired height preferably well above the soil surface, but where the stem diameter of the rootstock matches the stem diameter of the scion that has just been prepared. It is important to leave some foliage on the rootstock so that photosynthesis can continue. Using a sterilised blade or very sharp knife, carefully make a vertical cut down the centre of the rootstock stem (see diagram 2 on next page). The cut should be about the same length as the wedge section in the scion.

Place the wedge of the scion into the vertical cut of the rootstock (see diagram 3 on next page). Ideally there should be no section of the scion wedge exposed above the rootstock. Cuts can be easily modified to ensure a correct fit.

Make sure the graft is clean and dry, and tightly wrap the whole section from below the vertical cut to above the wedge using laboratory film or grafting tape (see diagram 4 on next page). This secures the graft and prevents water getting into the tissue. The tape will expand with the stem section as it grows, and eventually break down and fall off.

Materials necessary for grafting.

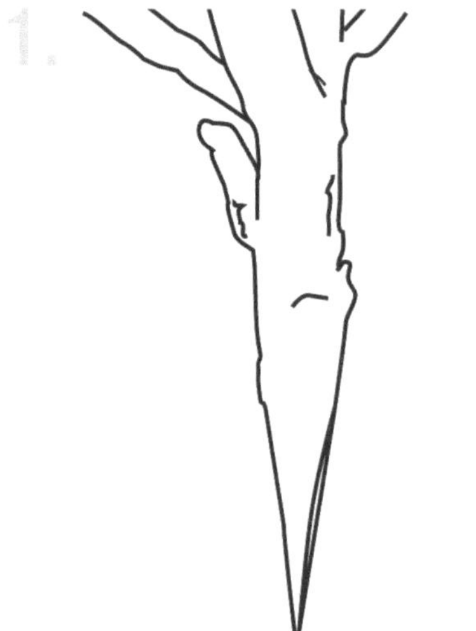

A scion with both sides cut ready for insertion into the rootstock.

Make a vertical cut into rootstock stem.

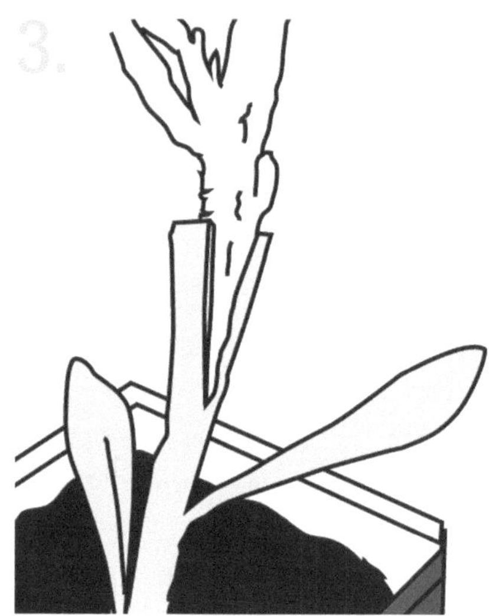

Insert the scion into the rootstock.

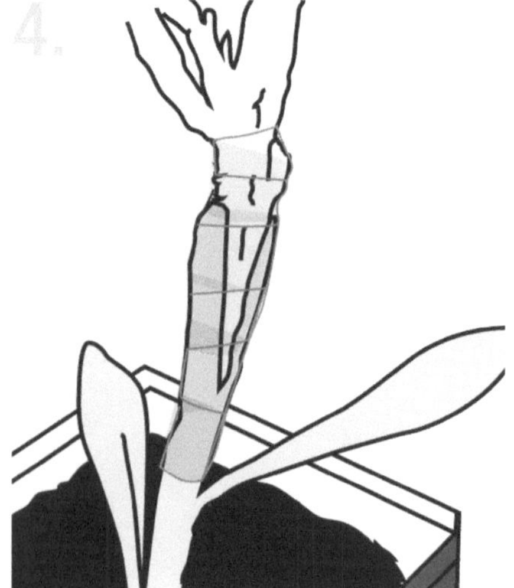

Tape the wedge graft up properly.

Completed grafts should be placed in a humid environment to prevent drying out. Keep the rootstock soil moist but not too wet, and take care to avoid excess water on the scion.

Some rootstocks are very vigorous and regular checking and pruning of the side shoots is necessary to prevent them overtaking the scion.

The time for a grafted plant to 'take' is dependent upon species, time of year and environmental conditions, but signs of success are usually evident within 4 to 6 weeks. Signs of a successful graft include callusing of the graft union and new growth of the scion. Once the graft shows obvious signs of success it can be gradually hardened off. Rootstock foliage may be completely pruned off at this stage but might require ongoing and regular removal for some time depending on the species.

Budding is a grafting technique regularly used in fruit and nut trees and some ornamentals but it has not been widely used on Australian native plants. Early results of trials at Kings Park on *Corymbia ficifolia* and *C. ptychocarpa* hybrids using the 'T' budding technique have been very promising. However, the long-term success of budding for many potential combinations of Australian plants is not known.

Table 3: Characteristics of selected rootstock tried in the Kings Park nursery

Rootstock	Strike rate	Growing on	Strengths	Weaknesses
Boronia clavata	Good	Quick	Compatible with a range of species; hardy	Susceptible to scale and mealybug
Boronia crenulata	Good	Quick	Thin diameter for smaller species	Susceptible to scale and mealybug
Chamelaucium × *Verticordia* 'Paddy's Pink'	Good	Quick	Thin diameter for smaller species	Highly susceptible to botrytis in glasshouse
Chamelaucium floriferum × *uncinatum*	Good	Quick	Good initial compatibility with a range of species	Possibly very susceptible to phytophthora
Chamelaucium × *Verticordia* var. 'Jasper'	Good	Quick	Thin diameter for smaller species	Highly susceptible to botrytis in glasshouse; poor vigour
Darwinia citriodora	Good	Quick	Phytophthora tolerant; grows in a range of conditions	Displays signs of nutrient deficiencies in alkaline soils
Grevillea preissii	Average	Moderate	Initial trials show good compatibility	Not generally long-lived
Grevillea hybrid	Good	Moderate	Quick strike	High losses in humid environment; susceptible to rotting
Myoporum insulare	Good	Quick	Thick diameter good for matching thicker species; hardy	Highly susceptible to aphids and scale
Myoporum montanum	Good	Quick	Nematode tolerant; good compatibility with a range of species	Susceptible to aphids and scale; high maintenance post-graft
Myoporum tetrandum	Good	Quick	Thin diameter; good compatibility with a range of species	Susceptible to nematodes, aphids and scale; high maintenance post-graft
Pimelea ferruginea	Good	Moderate	Good for coastal conditions	Relatively short-lived; not ideal for a wide range of soils
Westringia dampieri	Good	Quick	Good compatibility with a range of species; hardy	High maintenance post-graft

Table 4: Selected proven rootstock–scion combinations at Kings Park			
Scion	Rootstock	Success rate	Longest lived specimen
Eremophila nivea	*Myoporum montanum*	53%	10 years (Botanic Garden)
Microcorys eremophiloides	*Westringia dampieri*	45%	13 years (container stock); 6 years (Botanic Garden)
Pimelea ciliata	*Pimelea ferruginea*	86%	6 years (Botanic Garden; NB: trials on this species only began 6 years ago)
Pimelea physodes	*Pimelea ferruginea*	88%	7 years (Botanic Garden; NB: trials on this species only began 7 years ago)
Prostanthera magnifica	*Westringia dampieri*	56%	20 years (Botanic Garden); 17 years (container stock)

Table 5: Average number of days for cuttings to strike for selected species		
Species	Days	Comments
Acacia ashbyae	31	Variable strike rate; can be temperamental to grow on after potting; does not like root disturbance
Acacia forrestiana	32	Consistently good strike rate
Acacia lanuginophylla	28	Strikes well but often high mortality after potting
Acacia stenoptera	52	Best results in early summer
Acanthocarpus preissii	113	Difficult to strike
Actinodium cunninghamii	34	Consistently good strike rate
Adenanthos acanthophyllus	72	Slow to develop after potting
Adenanthos barbiger	42	Trickier than many other *Adenanthos* species; best results in late spring
Adenanthos cuneatus	24	Consistently good strike rate
Adenanthos cygnorum ssp. *cygnorum*	32	Best results from small, soft tips (limited success from harder material); timing and condition of material strongly influence results
Adenanthos cygnorum ssp. *chamaephyton*	33	Best results in late summer
Adenanthos detmoldii	22	Consistently good strike rate
Adenanthos linearis	37	Can be temperamental; quality of cutting is very important
Adenanthos obovatus	29	Consistently good strike rate
Adenanthos sericeus	32	Consistently good strike rate
Adriana quadripartita	42	Consistent strike rate; best results in spring to summer
Alyogyne huegelii	29	Consistently good strike rate
Alyogyne pinoniana	26	Consistently good strike rate
Alyogyne wrayae	69	Consistently good strike rate
Anthocercis intricata	39	Variable strike rates; best results mid to late summer
Anthotroche pannosa	40	Best results with rockwool
Anthotroche walcottii	45	Best results with rockwool
Astartea fascicularis	32	Consistently good strike rate
Astroloma ciliatum	99	Best results with tiny, soft tips in shallow cell trays, Clonex® 1500 ppm and humid environment
Astroloma foliosum	57	Best results with tiny, soft tips in shallow cell trays, Clonex® 1500 ppm and humid environment

Table 5: Average number of days for cuttings to strike for selected species (CONTINUED)

Species	Days	Comments
Astroloma macrocalyx	68	Best results with tiny, soft tips in shallow cell trays, Clonex® 1500 ppm and humid environment
Astroloma pallidum	77	Best results with tiny, soft tips in shallow cell trays, Clonex® 1500 ppm and humid environment
Babingtonia behrii	46	Best results in early spring
Babingtonia camphorosmae	90	Best results in late spring to early summer after flowering
Baeckea grandis	22	Strikes well in winter months
Baeckea latens	37	Reliable strike rate throughout the year
Banksia brownii	51	Best results in individual fibre pots
Banksia coccinea	73	Best results in individual fibre pots
Banksia nivea	36	Very difficult; low strike rate
Banksia nivea ssp. *uliginosa*	94	Very difficult; low strike rate
Beaufortia squarrosa	32	Best results from tip growth in spring to summer
Billardiera heterophylla	51	Strikes well in spring to summer
Boronia alata	69	Best results in late spring to summer
Boronia clavata	45	Strikes well throughout the year
Boronia crenulata ssp. *crenulata*	53	Best results in late spring to autumn
Boronia heterophylla	55	Best results with tiny, soft tips in shallow cell trays, Clonex® 1500 ppm and humid environment
Boronia megastigma	53	Best results in late spring after flowering
Calytrix acutifolia	28	Consistently good strike rate
Calytrix aurea	40	Best results in late spring to summer
Calytrix fraseri	37	Best results in late spring to summer
Calytrix tetragona	32	Consistently good strike rate
Chamelaucium axillare	24	Best results in late spring to summer after flowering
Chamelaucium ciliatum	34	Consistent strike rate
Chamelaucium megalopetalum	49	Generally good strike rate from tips or semi-hardwood
Chamelaucium uncinatum	38	Consistent strike rate
Cheyniana microphylla	37	Average strike rate in late spring to summer; can be temperamental
Chorilaena quercifolia	67	Best results with rockwool
Chorizema ilicifolium	21	Consistently good strike rate
Chorizema varium	55	Consistently good strike rate
Commersonia pulchella	36	Material quality very important
Conospermum boreale	74	Best results in individual fibre pots and with semi-hardwood
Conospermum stoechadis	48	New tip growth gives best strike rate, can also get results with semi-hardwood material
Conospermum triplinervium	72	Best results in individual fibre pots
Conospermum wycherleyi	78	Best results in individual fibre pots in summer
Conostephium pendulum	124	Difficult species; moderate results with small, soft tips
Coopernookia georgei	25	Good results in late winter to summer

Table 5: Average number of days for cuttings to strike for selected species (CONTINUED)		
Species	Days	Comments
Cyanostegia lanceolata	*59*	Best results with rockwool
Dampiera altissima	*42*	Best results with rockwool
Dampiera diversifolia	*29*	Consistently good strike rate
Dampiera hederacea	*43*	Good strike rate from semi-hardwood in spring to summer
Dampiera incana	*45*	Best results in late spring
Dampiera linearis	*48*	Best results with semi-hardwood
Dampiera teres	*33*	Consistently good strike rate
Dampiera trigona	*34*	Consistently good strike rate
Darwinia citriodora	*31*	Strikes consistently throughout the year
Darwinia ferricola	*29*	Strikes consistently throughout the year
Darwinia macrostegia	*47*	Reasonable strike rate
Darwinia meeboldii	*45*	Best results in late spring after flowering
Darwinia neildiana	*26*	Consistently good strike rate
Darwinia oldfieldii	*51*	Consistently good strike rate
Darwinia oxylepis	*44*	Best results in late spring after flowering
Darwinia squarrosa	*48*	Best results in late spring after flowering
Daviesia cunderdin	*57*	Can be tricky, but best results in individual fibre pots
Daviesia epiphyllum	*34*	Best results over spring/summer
Daviesia euphorbioides	*55*	Excellent results with rockwool; harder to strike in soil-based media
Diaspasis filifolia	*18*	Consistently good strike rate
Dicrastylis fulva	*43*	Best results in spring/summer; does well in rockwool or as small tips in cell trays with limited overhead watering
Dicrastylis incana	*59*	Best results in spring/summer; does well in rockwool or as small tips in cell trays with limited overhead watering
Diplolaena andrewsii	*44*	Best results with rockwool
Diplolaena angustifolia	*63*	Best results with rockwool
Diplolaena velutina	*29*	Best results with rockwool
Dodonaea ceratocarpa	*19*	Very quick to strike at consistently high rate
Eremophila brevifolia	*34*	Best results in early autumn
Eremophila glabra	*28*	Consistently good strike rate from tips as well as harder material
Eremophila nivea	*40*	Best results with rockwool
Eremophila platycalyx	*58*	Can be difficult to strike (best grafted)
Eremophila racemosa	*60*	Moderate results in spring
Eremophila splendens	*54*	Best results with rockwool
Frankenia hispidula	*50*	Best results in spring
Frankenia laxiflora	*33*	Can be temperamental
Geleznowia verrucosa	85	Moderate results in spring (best grafted)
Goodenia grandifolia	26	Best results in individual fibre pots and with semi-hardwood
Goodenia varia	24	Consistently high strike rate throughout the year
Grevillea armigera	47	Best results in individual fibre pots and with semi-hardwood

Table 5: Average number of days for cuttings to strike for selected species (CONTINUED)

Species	Days	Comments
Grevillea astericosa	52	Best results in individual fibre pots and with semi-hardwood
Grevillea baxteri	29	Best results in individual fibre pots and with semi-hardwood
Grevillea beardiana	50	Best results in individual fibre pots and with semi-hardwood
Grevillea bipinnatifida ssp. *bipinnatifida*	49	Best results in individual fibre pots and with semi-hardwood
Grevillea bipinnatifida ssp. *pagna*	73	Best results in individual fibre pots and with semi-hardwood
Grevillea bracteosa	50	Best results in individual fibre pots and with semi-hardwood
Grevillea concinna ssp. *lehmanniana*	64	Best results in individual fibre pots and with semi-hardwood; can be temperamental
Grevillea curviloba ssp. *incurva*	42	Consistently good strike rate from tips as well as harder material
Grevillea dielsiana	70	Best results in individual fibre pots and with semi-hardwood
Grevillea drummondii	35	Consistently good strike rate from tips as well as harder material
Grevillea dryandroides ssp. *dryandroides*	67	Best results in individual fibre pots and with semi-hardwood
Grevillea dryandroides ssp. *hirsuta*	71	Best results in individual fibre pots and with semi-hardwood
Grevillea endlicheriana	35	Best results in individual fibre pots and with semi-hardwood
Grevillea fililoba	38	Consistently good strike rate from tips as well as harder material
Grevillea flexuosa	32	Best results in individual fibre pots and with semi-hardwood
Grevillea hookeriana	63	Best results in individual fibre pots and with semi-hardwood; can be temperamental
Grevillea huegelii	49	Best results in individual fibre pots and with semi-hardwood
Grevillea humifusa	37	Consistently good strike rate from tips as well as harder material
Grevillea insignis ssp. *elliotii*	53	Best results in individual fibre pots and with semi-hardwood
Grevillea insignis ssp. *insignis*	50	Best results in individual fibre pots and with semi-hardwood
Grevillea involuctrata	76	Best results in individual fibre pots and with semi-hardwood
Grevillea maccutcheonii	40	Best results in individual fibre pots and with semi-hardwood
Grevillea maxwellii	40	Best results in individual fibre pots and with semi-hardwood
Grevillea patentiloba ssp. *platypoda*	33	Consistently good strike rate from tips as well as harder material
Grevillea petrophiloides ssp. *magnifica*	65	Best results in individual fibre pots and with semi-hardwood
Grevillea petrophiloides ssp. *petrophiloides*	97	Best results in individual fibre pots and with semi-hardwood
Grevillea pimeleoides	32	Consistently good strike rate from tips as well as semi-hardwood
Grevillea preissii ssp. *glabrilimba*	48	Consistently good strike rate from tips as well as harder material
Grevillea preissii ssp. *preissii*	51	Best results using very small, soft tips in shallow cell trays and humid environment
Grevillea prostrata	27	Consistently good strike rate from tips as well as harder material
Grevillea saccata	49	Best results in individual fibre pots and with semi-hardwood
Grevillea stenomera	35	Best results in individual fibre pots and with semi-hardwood
Grevillea synapheae ssp. *synapheae*	34	Consistently good strike rate from tips as well as harder material
Grevillea thyrsoides ssp. *pustulata*	25	Best results in individual fibre pots and with semi-hardwood

Table 5: **Average number of days for cuttings to strike for selected species** (CONTINUED)

Species	Days	Comments
Grevillea tripartita ssp. *macrostylis*	68	Best results in individual fibre pots and with semi-hardwood
Grevillea vestita	54	Best results in individual fibre pots and with semi-hardwood
Hakea myrtoides	54	Best results in individual fibre pots and with semi-hardwood
Hemiandra pungens	38	Excellent strike rate throughout the year
Hibbertia acerosa	42	Best results from small, soft tips in shallow cell trays
Hibbertia aurea	57	Best results from small, soft tips in shallow cell trays
Hibbertia grossulariifolia	40	Consistently good strike rate
Hibbertia huegelii	66	Moderate strike rate
Hibbertia hypericoides	47	Best results from small, soft tips in shallow cell trays in spring to summer
Hibbertia racemosa	40	Best results in late spring to summer
Hibbertia stellaris	33	Consistently good strike rate
Hibbertia subvaginata	33	Consistently good strike rate
Homalocalyx thryptomenoides	36	Good results in spring
Hybanthus calycinus	77	Best results from small, soft tips in high humidity environment
Hypocalymma angustifolium	30	Consistently good strike rate
Hypocalymma cordifolium	33	Consistently good strike rate
Hypocalymma robustum	47	Consistently good strike rate
Hypocalymma strictum	47	Consistently good strike rate
Hypocalymma tetrapterum	30	Consistently good strike rate
Hypocalymma xanthopetalum	33	Consistently good strike rate
Isopogon cuneifolia	37	Best results in spring to summer after flowering
Isopogon dubius	41	Good results in individual fibre pots and with semi-hardwood
Isopogon formosus	91	Good results in individual fibre pots and with semi-hardwood
Isopogon trilobus	49	Good results in individual fibre pots and with semi-hardwood
Jacksonia velveta	34	Best results in individual fibre pots; can be temperamental
Keraudrenia hermanniifolia	55	Can be difficult
Keraudrenia integrifolia	40	Can be difficult
Keraudrenia velutina ssp. *velutina*	29	Can be difficult
Kunzea cincinnata	40	Best results in spring
Kunzea pauciflora	36	Good strike rate throughout the year
Kunzea pulchella	29	Best results in spring to summer
Lachnostachys verbasifolia	61	Best results in rockwool
Lambertia inermis	66	Best results in individual fibre pots and with semi-hardwood
Lambertia orbifolia	46	Best results in individual fibre pots and with semi-hardwood
Lambertia rariflora	52	Best results in individual fibre pots and with semi-hardwood
Lasiopetalum membranaceum	57	Strikes consistently well from soft and semi-hardwood material
Lasiopetalum pterocarpum	64	Strikes consistently well from soft and semi-hardwood material
Lechenaultia biloba	33	Strikes consistently well from soft and semi-hardwood material
Lechenaultia chlorantha	30	Strikes consistently well from soft and semi-hardwood material
Lechenaultia expansa	36	Strikes consistently well from soft and semi-hardwood material

Table 5: Average number of days for cuttings to strike for selected species (CONTINUED)

Species	Days	Comments
Lechenaultia floribunda	37	Strikes consistently well from soft and semi-hardwood material
Lechenaultia formosa	43	Strikes consistently well from soft and semi-hardwood material
Lechenaultia linarioides	38	Strikes consistently well from soft and semi-hardwood material
Lechenaultia macrantha	37	Strikes well but can be tricky to grow on
Leptosema aphyllum	42	Best results in late spring to early autumn
Leptospermum sericeum	36	Best results in spring
Leucophyta brownii	34	Good strike rate but needs very well drained propagation media
Leucopogon gnaphalioides	72	Best results in individual fibre pots and with semi-hardwood
Leucopogon parviflorus	102	Moderate strike rate in summer but very slow root development
Leucopogon propinquus	140	Very difficult; low strike rate
Lobelia anceps	38	Good strike rate throughout the year
Lysiosepalum involucratum	49	Best results in spring to summer
Maireana sedifolia	40	Reasonable strike rate throughout the year but can be difficult to grow on over cooler months
Marianthus bicolor	33	Can be temperamental (best from seed)
Marianthus paralius	50	Best results in late spring to summer
Melaleuca fulgens ssp. *fulgens*	28	Consistently high strike rate throughout the year
Melaleuca fulgens ssp. *steedmanii*	18	Consistently high strike rate throughout the year
Microcorys eremophiloides	56	Can be tricky (best grafted)
Myoporum insulare	38	Best results with rockwool
Myoporum montanum	64	Best results with rockwool
Nematolepis phebalioides	53	Best results in late winter to summer
Newcastelia insignis	51	Best results with rockwool
Newcastelia roseoazurea	51	Best results with rockwool
Nuytsia floribunda	72	Very difficult from cuttings
Olearia axillaris	33	Best results using a free-draining mix with organic component and slow release fertiliser
Olearia occidentissima	27	Best results in late spring to summer
Opercularia vaginata	28	Can be temperamental
Persoonia longifolia	47	Can be tricky (best results in winter months)
Persoonia micranthera	113	Best results in individual fibre pots with fresh new growth
Petrophile biloba	41	Best results in individual fibre pots and with semi-hardwood
Petrophile chrysantha	37	Best results in late spring
Phebalium canaliculatum	88	Can be temperamental (best grafted)
Philotheca spicata	62	Best results in late spring
Physopsis chrysophylla	65	Best results in late spring and summer, with small tips in cell trays with no overhead watering
Pileanthus filifolius	63	Material quality is an important factor
Pileanthus peduncularis ssp. *peduncularis*	31	Best results in early summer
Pimelea argentea	48	Best results with rockwool

Table 5: Average number of days for cuttings to strike for selected species (CONTINUED)

Species	Days	Comments
Pimelea clavata	62	Can be tricky; best results using softer material
Pimelea ferruginea	50	Consistently high strike rate throughout the year
Pimelea rosea	57	Best results in late spring to summer
Pimelea suaveolens	67	Can be tricky (best grafted)
Pimelea sylvestris	49	Best results with rockwool
Pityrodia atriplicina	61	Best results in summer
Pityrodia dilatata	50	Best results in summer
Pityrodia terminalis	49	Best results in summer
Pityrodia verbascina	82	Best results with rockwool
Platysace compressa	56	Best results in spring
Platytheca galioides	35	Best results in spring to summer
Prostanthera magnifica	48	Generally low strike rate (best grafted)
Prostanthera striatiflora	50	Best results in summer
Ptilotus obovatus	29	Best results using a free-draining mix with organic component, slow release fertiliser and limited overhead watering
Ptilotus polakii	33	Best results using a free-draining mix with organic component, slow release fertiliser and limited overhead watering
Regelia velutina	46	Best results from small, soft tips in shallow cell trays
Rulingia densiflora	42	Best results from semi-hardwood
Rulingia kempeana	28	Best results from semi-hardwood in late spring to summer
Sarcostemma viminale ssp. *australe*	34	Best results in late spring to summer, with care taken not to overwater
Scaevola aemula	25	Consistently high strike rate throughout the year
Scaevola anchusifolia	45	Best results in spring to summer
Scaevola calliptera	38	Best results in individual fibre pots and with semi-hardwood
Scaevola crassifolia	39	Best results in spring to summer
Scaevola nitida	55	Best results in spring to summer
Scaevola oldfieldii	44	Best results in spring to summer
Scaevola platyphylla	48	Best results in individual fibre pots and with semi-hardwood
Scaevola spinescens	56	Best results in spring to summer
Scholtzia laxiflora	34	Best results in late spring to summer after flowering
Scholtzia parviflora	41	Best results in summer after flowering
Spinifex longifolius	20	Best results in spring to summer and in individual pots
Stirlingia latifolia	88	Best results in individual fibre pots or rockwool
Stylidium adnatum	30	Best results in late summer to autumn when new growth starts
Stylidium brunonianum	21	Best results in spring
Stylidium bulbiferum	34	Best results in spring
Stylidium repens	63	Best results in winter to spring
Tecticornia verrucosa	49	Best results in late spring to summer; do not overwater
Tetratheca hirsuta	32	Best results in late spring to autumn
Thomasia foliosa	39	Best results in spring to summer

Table 5: Average number of days for cuttings to strike for selected species (CONTINUED)

Species	Days	Comments
Thomasia glabripetala	45	Best results in spring to summer
Thomasia macrocarpa	62	Best results with rockwool
Thomasia purpurea	39	Consistently good strike rate throughout the year
Thomasia quercifolia	51	Best results in spring to autumn
Thryptomene baeckeacea	43	Best results in spring to summer after flowering
Thryptomene denticulata	26	Consistently good strike rate throughout the year
Thryptomene hyporhytis	29	Consistently good strike rate throughout the year
Thryptomene saxicola	41	Consistently good strike rate throughout the year
Thryptomene stenophylla	31	Best results in spring to summer after flowering
Thryptomene strongylophylla	37	Best results in spring to summer after flowering
Tremandra stelligera	29	Best results in late spring to summer
Velleia foliosa	32	Best results in spring to summer
Verticordia argentea	37	Difficult to achieve high strike rate (best grafted)
Verticordia attenuata	31	High strike rate throughout the year with good quality material
Verticordia brownii	38	Best results in late spring to summer
Verticordia chrysantha	30	High strike rate throughout the year with good quality material
Verticordia chrysanthella	28	High strike rate throughout the year with good quality material
Verticordia citrella	42	High strike rate throughout the year with good quality material
Verticordia cooloomia	39	Best results after flowering when new growth is produced
Verticordia cunninghamii	44	Best results in summer
Verticordia eriocephala	44	High strike rate throughout the year with good quality material
Verticordia etheliana var. *etheliana*	32	Difficult to achieve high strike rate (best grafted)
Verticordia etheliana var. *formosa*	64	Difficult to achieve high strike rate (best grafted)
Verticordia fastigiata	60	Best results in summer
Verticordia forrestii	38	Difficult to achieve high strike rate (best grafted)
Verticordia galeata	40	High strike rate throughout the year with good quality material
Verticordia grandis	45	Difficult to achieve high strike rate (best grafted)
Verticordia helichrysantha	31	High strike rate throughout the year with good quality material
Verticordia huegelii var. *huegelii*	22	Best results in spring
Verticordia mitchelliana	35	High strike rate throughout the year with good quality material
Verticordia monadelpha var. *callitricha*	57	High strike rate throughout the year with good quality material
Verticordia monadelpha var. *monadelpha*	51	High strike rate throughout the year with good quality material
Verticordia oculata	36	Difficult to achieve high strike rate (best grafted)
Verticordia ovalifolia	42	Best results in spring
Verticordia plumosa var. *plumosa*	31	High strike rate throughout the year with good quality material
Westringia dampieri	43	Good strike rate with soft tips of semi-hardwood
Xanthosia rotundifolia	46	Best results in late spring to summer

Chapter 6

Seeds

Luke Sweedman and David Merritt

Seeds of wild species can be used as a cost-effective and simple means of producing plants for the horticultural industry, conservation programs and restoration of damaged landscapes. When correctly stored in seed banks, seeds also represent an important, secure resource for long-term biodiversity conservation. Most seeds can be stored for many years, decades or even centuries, and still be propagated readily.

There is a rapidly increasing demand for seeds of wild species. Most seeds from Australian native plants are collected from the wild, placing a responsibility on the seed collector to be efficient, careful and sustainable in obtaining seeds from what is often a diminishing resource. As more land is cleared or impacted by drought, weeds or other disturbances, seeds become more difficult to find and the impact of their removal from the wild becomes greater. Seeds collected from wild plant populations are highly valuable both ecologically and economically, and every effort must be made to use seeds ethically.

Seeds are relatively easy to collect from most plant species. For conservation purposes, collecting seeds correctly will provide a good genetic representation of the population. However, some plants produce seeds only sparingly and, for these, careful planning is required to ensure collecting expeditions are successful. Also, it is important to note that not all desirable plant characteristics may necessarily be passed on through the seeds. Some particular features, such as a specific flower or foliage colour, may be better reproduced through propagation via cuttings rather than from seeds. The popular red-flowering gum (*Corymbia ficifolia*), for example, will provide a variety of colour forms when propagated from seeds. Also, some forms of plants which have certain appealing habits, such as a prostrate form, may be produced through environmental influences and not inherited by the seeds. Thus a good working knowledge of a species can be important if targeting plants for a specific outcome.

Seeds are one of the few biological organisms able to survive extreme desiccation. Most seeds comprise only 20 to 30 per cent water when fully mature, and can survive further desiccation to less than 5 per cent water content. This remarkable ability to lose almost all cellular water and still survive means that many seeds can be stored for long periods. Desiccation-tolerant seeds are termed 'orthodox' and most seeds from temperate, Mediterranean and arid climates are orthodox. Because they can be dried to a low water content,

Luke Sweedman

A seed collector may find new and interesting forms of plants created by local environmental conditions, such as this dwarf form of *Melaleuca cardiophylla* collected near Seabird, Western Australia. Only by germinating the collected seeds and growing the resultant plants under horticultural conditions, can it be seen if the desirable form has been inherited by the seeds.

orthodox seeds can be stored at freezing temperatures without the risk of lethal ice crystals forming within the cells. It is this ability to dry seeds and then store them at freezing temperatures that enables seed banks to conserve seeds for long periods.

A smaller group of seeds, termed 'recalcitrant', do not survive water loss. These seeds are typically produced by plants from tropical environments that are wet throughout the year. Recalcitrant seeds are difficult to store because they cannot be dried or frozen. But the vast majority of Australian native seeds can be dried and stored following simple procedures.

Luke Sweedman

The closed valves of the serotinous fruits of *Eucalyptus macrocarpa* are clearly visible. Seeds can be retained for several years within these protective, woody fruits.

Seed collecting

For many species, seed collecting is straightforward, requiring simple tools and little cost. However, before you start collecting seeds, the purpose of the collection and the length of time the seeds will be stored should always be determined. Specific and rigorous protocols apply to seeds collected for long-term conservation. Seeds to be used to generate plants for horticultural display can often be sourced from cultivated plants growing in domestic parks and gardens. In such cases seeds can simply be collected with a minimum of information and without any specific sampling requirements. Collecting seeds from wild plant populations requires more rigour in terms of plant identification, record-keeping, sampling regimes and protection of the source populations. Maintaining the genetic integrity of a seed collection is crucial when collecting for use in provenance-specific restoration plantings that reflect a particular location or habitat.

The best results from seed collection expeditions will always be obtained when a good understanding and knowledge of the target species is first developed. For example, within the large genus *Eucalyptus* the majority of species retain fruits for several years, making it easy to collect them at most times of the year. However, the *Eucalyptus* species from northern Australia generally do not retain seeds but, rather, flower during the wet season, produce fruit and then drop the seeds as they mature. A visit to these species at the wrong time of year may be an expensive waste of resources.

Collecting seeds of many species requires careful observation and timing. This may entail several visits to the plants from which the seeds are to be collected. It is a good strategy to collect samples of the target species while they are in flower, in order to verify the species identification and the plant form or habit. Pressed specimens with unique collection numbers assigned, together with details regarding the flowering and fruiting times, are a valuable asset as a field guide and herbarium record. Future seed collection trips can then be judged with more accuracy.

Some plants produce staggering amounts of seeds. Species of *Melaleuca*, for example, can produce millions of seeds in each flowering cycle. An individual coastal tea tree (*Melaleuca lanceolata*) plant can hold this quantity of seed for its entire life unless a disturbance affects it. Conversely, in some plants seed production is scarce. Cycads may produce only a few hundred seeds over a number of years and some banksias produce very little seed each year, with those that are produced often heavily predated by insects. Seed production can be maintained by the plant for its whole life and seed collection can be a highly sustainable activity if it is limited and controlled through careful assessment of the plant population, seed availability and the seed storage requirements. Collecting from as many plants as practicable in a wild population ensures good genetic representation and spreads the impact of collecting. Populations harvested in this fashion can be visited annually without a decrease in seed production or population vigour. Alternatively, seed from one garden plant may supply a very good source from which to propagate material for horticultural use.

This Kings Park seed collecting expedition involved spending several weeks in the remote Kimberley region of Western Australia.

An appropriate and ethical sampling strategy is to take only the required seed material from individual plants in a way that minimises the physical impact, maximises the range of genetic diversity captured and ensures sufficient seeds remain available for future natural regeneration.

Fruits or seeds collected from within a single population of plants should be collected from a number of randomly sampled plants. A population may contain anything from a few individual plants to many thousands. Seeds collected from plants chosen randomly will provide a wider sampling of the genetic diversity, as opposed to seeds collected from only one plant. The exact number of plants to be sampled is dependent on the proposed end use of the seeds, the size of the population and, to some extent, time and practical constraints. If collecting to capture genetic representation from different habitats across a landscape, it is wise to keep each seed collection separate. In this way seeds from specific areas can be planted back into similar habitats. Sampling rare species will require keeping seeds from individual plants separate. In most cases a random collection from at least 10 plants within a population of several hundred will suffice.

Notes for seed collecting

- Collect according to protocols appropriate to the intended use of the seeds (for example, short-term storage for restoration or horticultural plantings, or long-term storage for conservation).
- Many plants produce large numbers of seeds. Collect only the seeds you need. It is recommended to collect only 20 per cent of the available seed from any single plant.
- For cultivated garden plants keep seeds of each species separate, and be sure to record basic information.
- For wild populations ensure collections are made in an environmentally sensitive and sustainable manner.
- For wild species keep all seed collections separate, and record detailed information pertaining to site characteristics. Take herbarium samples to ensure correct identification.
- Always collect seeds when mature. Immature seeds may not store well.

Before leaving on a collection trip, plan what you want to collect and where the species can be located. If collecting from wild populations, make sure that you have the appropriate permission or licences. Seeds of Australian plants are most often collected between late spring and summer in the temperate climatic regions, or in autumn and winter in the northern, summer rainfall zones. Spring flowering plants will complete flowering after several weeks, then produce fruits that develop, ripen and produce mature seeds over a period of weeks or months.

About two-thirds of Australian plant species are geosporous, meaning they release mature seeds into the soil. For these species, successful seed collecting depends on correct timing. The natural dispersal of seeds may take place over just a few days or over many weeks, so the skill is to be in the right place at the right time to collect seeds at the point at which they are mature, but before they begin to disperse. Visiting plant populations regularly may be necessary to understand the timing of seed maturation and release.

The widespread daisy family (Asteraceae) comprises a large number of geosporous species that require careful timing for successful collection. Species within the Asteraceae are wind-pollinated and the seeds are wind-dispersed, being released quickly over just a few days. In Western Australia, flowers of most species are at their peak in late August and seeds are ready to collect by mid-September. The seeds of some species that flower later are collected into the summer months.

Seeds from Asteraceae plants need to be collected as they are starting to dehisce and disperse. The perfect time to collect is when the whole population is about 80 per cent mature. But a single windy day can send these seeds many kilometres away, so planning and timing are paramount.

Luke Sweedman

The seeds of Asteraceae species are dispersed by the wind. Timing is everything when collecting these seeds as they can literally be blown away overnight.

The wattles (*Acacia*) are another common group of geosporous species. The genus *Acacia* is the largest plant group in Australia. Seeds of *Acacia* are usually very easy to collect, with all species producing seedpods about 6 to 8 weeks after flowering. While these seedpods vary enormously in size, shape and form, they should all ideally be collected when dry and when the seeds are dehiscing from the pods. Seeds will be hard, usually lustrous, and well filled. While *Acacia* seeds can be collected when the fruits are still slightly immature (as long as they are fully formed) seeds of most species will not store well if collected at this point. As a last resort, immature seeds of some species may still be successfully matured after collection, provided that sections of the stems with the attached fruits are removed from the plant and these sections are then kept under ambient conditions for several weeks to allow some moisture to be retained within the fruits. But there is no substitute for collecting mature, naturally dehiscing seed.

About one-third of Australian plants are serotinous, meaning they hold their mature seeds in woody fruits in the plant canopy, rather than immediately dispersing the seeds. In the wild, serotinous fruits release their seeds when the plants are subject to a fire, stress or other disturbance. Serotinous species can be collected from the plants at most times of the year, but collecting during the flowering period is ideal as plant identification is easier and the fruits are at least one year old. Fruits should be at least one year old to ensure the seeds are

Luke Sweedman

Mature, ready-to-collect pods of (from left to right) *Acacia aneura, Abrus precatorius, Pittosporum phylliraeoides* and *Grevillea stenobotrya*.

fully mature. This is especially important for most *Banksia, Beaufortia, Callistemon, Calothamnus, Eucalyptus, Hakea* and *Melaleuca* species. The woody fruits of many serotinous species will open naturally over a few days to weeks when they are removed from the plant. However, some fruits may need heating or repeated wetting and drying in order to open, especially those of *Banksia* species.

Luke Sweedman

The serotinous fruits of *Melaleuca huegelii*, ready to collect.

While collecting seeds, record keeping needs to be only as complex as required for the chosen use of the seeds. Perhaps the most important step is to collect plant voucher specimens to be used as a herbarium reference for verifying the identity of the collected species. Recording the species identification, collection location, associated vegetation and other relevant information is useful, especially for provenance-specific collections. Seed collections for commercial purposes should have a high level of information provided, including the precise collection location recorded as Global Positioning System (GPS) co-ordinates or directions from the nearest known town or named place, and additional seed quality information such as the percentage of viable seeds in the seed lot. A seed collection from a cultivated garden plant may only require the name of the plant to be recorded if the seeds are just needed to grow new plants for the garden.

For conservation or restoration purposes, however, rigorous records should be kept, including species identification verified by a botanist; the collection site details including habitat, associated vegetation, soil type, aspect, elevation, population size and condition, and GPS location; and a unique collection number. A database with a complete record inventory for all collections is the ideal system for large seed collections. Collecting books form excellent templates to record all collection information and these can be sourced from most seed firms.

BOTANIC GARDENS AND PARKS AUTHORITY

Collector.................... Field No.................... Date....................
Collected with..
Material Collected: Seed / Cutt. / Herb. Spec / Photo / Other....................
Field ID..
Current Name..
Determinant..
Nearest Named Place..
Location..
....
Lat S.................... Long E....................
Datum: GDA 94 / Other.................... Altitude....................
Plant Description: Erect / Prostrate / Compact / Open / Woody
Herb / Shrub / Tree / Annual / Succulent
Height (m).................... Flowers....................
Habitat Open / Low / Tall / Forest / Woodland / Shrubland / Heath
Grassland / Kwongan / Swampy / Coastal / Other....................
Vegetaion: Mallee / Mulga / Banksia / Casuarina / Spinifex / Samphire
Saltbush / Bluebush / Other....................
Associated Taxa..
....
Soils: White / Yellow / Red / Brown / Grey / Black
Sand / Clay / Loam / Peat / Lateritic
Geology: Laterite / Dolerite / Limestone / Ironstone / Sandstone
Granite / Other....................
Geomorphology: Outcrop / Ridge / Breakaway / Dune / Plain / Swale
Gully / Swamp / Riverbank / Lake edge / Salt lake
Clay pan / Other....................
No. Sampled.................... Pop Size....................
Comments..
....

The field book used by Kings Park seed collectors captures all of the information required for accurate record-keeping and formal accessioning of each seed batch.

Notes for planning a collecting trip

- Plan a collection strategy for each target species that includes the locations of the populations, the method(s) of seed collection and the timing of the collection, based upon when seeds will be mature.
- Determine the quantity of seed required and the purpose of the seed collection.
- Assemble the tools required for collection. These may include secateurs, power saws or hand-operated pole pruners, tags, marker pens, a record book, plant press and plant identification guides.
- Ensure appropriate bags or containers are available in which to place the collected material. These may be calico, hessian or paper bags of varying sizes. Plastic collapsible bins are ideal.
- Ensure all permits are obtained and that a GPS, maps and precise locations are available.
- Ensure the vehicle is appropriate and safe for the journey.

Luke Sweedman

Collecting bags, bins and secateurs are essential basic equipment for any seed collecting expedition.

Seed cleaning

Seed cleaning should take place as soon as the material is evenly dry and the fruits have released all of the seeds. Before cleaning, it is important to check that the seeds are viable. This is best done by taking a small sample, dissecting the seed and inspecting the seed tissues under a magnifier. Viable seeds should have a healthy embryo and (in some seeds) a healthy endosperm. Healthy seed tissues appear turgid and, in most cases, white or pale yellow in colour. If no viable seeds can be found, consider rejecting the seed lot before cleaning and storage.

A seed cleaning area should be open and well ventilated, with plenty of bench space, a good set of sieves and, in some cases, more specialised equipment. An outside, covered area is ideal and all health and safety equipment needs to be available, especially filter particle masks. Generally, mechanised cleaning machines are only required for larger amounts of seeds, and small seed lots can be cleaned by hand. The majority of plant species can be cleaned effectively and efficiently by using a range of sieves. Sieves are readily available and relatively cheap, and a set of 10 with aperture sizes ranging from 0.25 to 2.5 millimetres is ideal. The seed cleaning process should aim to separate the packing material (or chaff) from within the fruits, along with any remaining leaf, twig or non-seed material.

Luke Sweedman

A range of sieves can be used to effectively clean small batches of seeds.

Tips for cleaning seed of common Australian species

- **Myrtaceae (including *Eucalyptus* and *Melaleuca*):** Dry the collected fruits by placing in trays or cloth bags for 1 to 3 weeks and leaving in a well-ventilated area. Check that the fruit valves have opened and seeds are released. Separate seeds from chaff using a range of sieves.
- **Fabaceae (including *Acacia* and *Senna*):** Make sure that the seed collection is dried well for several weeks, especially if the pods are still slightly immature. Seeds of some species will naturally dehisce from the pods, whereas others are tightly held and will need to be manually removed from the pods. Sieve out any seeds already separated from the pods and set aside. If seeds can be separated by hand from the pods in small quantities, then do so. For larger amounts use a threshing machine. Once material has been threshed, separate seeds from chaff using an aspirated blower.
- **Asteraceae and other annuals:** Seeds can be readily cleaned by sieving and removing dead flower parts. This may require a winnowing machine or an aspirated blower to successfully complete.

Useful equipment for seed cleaning

- A range of sieves of varying aperture sizes
- Threshing machine
- Clipper or aspirated blowers
- A variety of plastic bins, brooms and cleaning brushes
- Compressed air for cleaning equipment.

Luke Sweedman

After a short period in dry conditions the valves of the woody *Eucalyptus erythrocorys* fruits open to release the seeds.

Insect control

Most insect damage to seeds occurs prior to collection. Many seeds may have been predated while still on the parent plants. This predation may continue after the fruits and seeds are collected. All seeds should be examined carefully for signs of insect predation. Grubs are common in some species. Freezing seeds (once appropriately dried) is the best way of destroying insects quickly and effectively. Alternatively, a pyrethrin-based insecticide can be sprayed onto the seeds when spread out on trays.

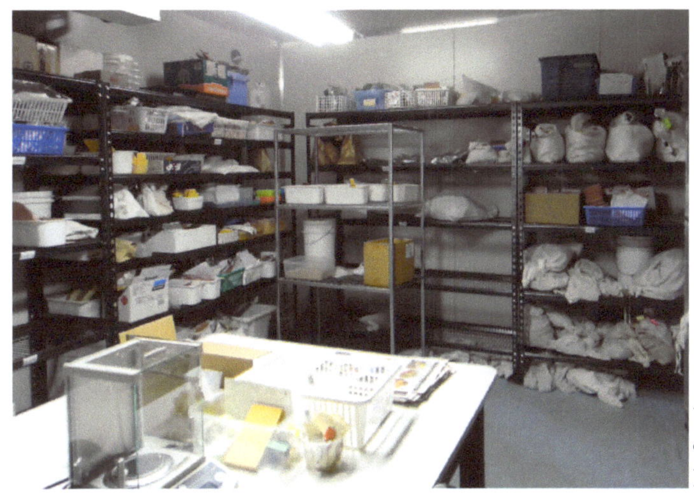
Luke Sweedman

The seed drying room at Kings Park.

Seed drying and storage

Post-harvest drying of seeds begins from the time that the fruits and seeds are removed from the plants in the field and it continues through the cleaning and storage process. Seed drying is the single most important factor influencing the longevity of the seed collection. A decrease in seed moisture content of just one per cent can double the storage life of most seeds. Drying seeds also facilitates the opening of woody fruits (for example, the capsules of *Melaleuca* species). It is important to place seeds into a drying environment as soon as possible after collection.

A drying environment can be simply a protected area that prevents seeds from taking up moisture from overnight dew and higher humidity. More sophisticated means of drying involve placing seeds in purpose-built drying rooms that control temperature and relative humidity. This is especially important for long-term storage, where seeds are ideally dried at between 15 to 20 per cent relative humidity at a temperature of 15 to 20°C. When drying seeds, it is important to spread them out evenly in an open container. If space is restricted, seeds can also be dried in calico or paper bags provided there is space for air circulation within the bags.

The seed drying room at Kings Park is maintained at 15°C and 15 per cent relative humidity. Seeds are kept in calico bags inside this controlled environment prior to cleaning. Once seeds are cleaned they are placed in paper packets for up to 4 weeks, then packaged in air-tight, laminated, aluminium foil bags and stored at –18°C.

Seed drying can also take place out in the open on tarpaulins or sheets, or in a well-ventilated, cool place if resources and facilities are limited or the seed collection is for the home garden. Seed drying usually takes a few weeks depending on the size of the seeds, the temperature and relative humidity and the seed moisture content upon initial collection. Seeds will dry fairly rapidly in temperatures of 25 to 30°C, and these temperatures are suitable for basic seed drying. Enclosed tunnel houses are also often used to dry seeds. These protect seeds from increased night-time humidity, but need to be well ventilated as the temperature inside can become extreme during the day.

Luke Sweedman

The fruits of serotinous species are often left outdoors to facilitate seed release before cleaning and drying.

The conditions under which seeds are stored vary depending on the purpose of the seed collection and the resources and technology available. At modern conservation seed banks seeds are dried under carefully controlled conditions (15 per cent relative humidity and 15°C), sealed in air-tight containers and then stored in freezers at −18°C. Under these conditions seeds of many Australian species are predicted to live for many hundreds, if not thousands, of years. For a shorter duration seeds can be stored in a fridge at 5°C, or even in an air-conditioned room, and these storage conditions are perfectly adequate for the home gardener. What is most important is to avoid prolonged exposure of seeds to high temperatures. For most species a reduction in temperature of 5°C will double the storage life. The choice of storage container is dependent upon the space available and the intended length of storage. Ideally, air-tight containers are used to prevent re-absorption of moisture by seeds following drying. Glass containers with rubber-sealed lids or hermetically sealed, laminated, aluminium foil bags are commonly used for storing seeds.

Seed ecology

A little knowledge of the ecology of seeds can greatly increase the chances of successful storage and germination. Understanding how seeds behave in the environment not only assists in seed collection but can also provide clues to germination and storage requirements.

Seeds provide a reproductive advantage to plants as they can be dispersed by wind and wildlife and last a long time before germinating, avoiding periods when the environment is unfavourable to plant growth. Seeds have many adaptations that allow their survival and persistence in soil (or the plant canopy) through sometimes many years of harsh extremes of climate. Seeds constantly sense and adapt to the surrounding conditions. Seeds are very sensitive to changes in temperature, soil moisture and light, and are able to detect and respond to minute concentrations of chemical agents in the soil, such as those released in smoke during a fire. In order to prevent germination in unsuitable environments for plant growth (such as during periods of seasonal drought), seeds can lie in the soil

Luke Sweedman

In the soil, seeds are constantly sensing the environment and only germinate at a time when the chances of seedling survival are greatest. Here, a flush of seedlings emerges following a fire in a jarrah (*Eucalyptus marginata*) forest.

in a dormant state. Seed dormancy prevents germination from occurring at the wrong time of year, during a drought-breaking rain-shower, for example, when there may be a transient period when the environmental conditions are suitable for plant growth, but the chances of seedling establishment are low. Seeds become non-dormant and able to germinate once the appropriate sequence of cues (normally changes in temperature and soil moisture) has been realised.

Seeds can be challenging to germinate, as the treatments suitable to break dormancy must first be applied and the incubation conditions must then be suitable to allow germination to proceed. Seed dormancy is complex as there are various means by which dormancy is imposed. Most seeds that are released into the soil at maturity (these plants are geosporous) are dormant. Dormancy can be imposed by the seed or fruit coat, by the embryo, or by a combination of both.

A seed coat that is impermeable to water is a common cause of dormancy, with most wattles (*Acacia* spp.) possessing this type of dormancy together with other species in the Convolvulaceae, Fabaceae, Malvaceae, Rhamnaceae and Sapindaceae. Hot water treatment or the physical abrasion (scarification) of the seed coat are artificial treatments commonly used to overcome this type of dormancy. Embryo-imposed dormancy is caused by a physiological 'block' to embryo growth that must be overcome before germination can proceed. Seeds with embryo-induced dormancy can become non-dormant following lengthy periods of storage under warm, dry conditions (after ripening), or following periods of exposure to warm, moist conditions (stratification) or repeated wetting and drying cycles. Treatment with smoke water or the plant growth regulator gibberellic acid can also promote germination of these types of seeds. Embryo-induced dormancy is very common, and occurs in families such as the Haemodoraceae, Myrtaceae, Poaceae, Proteaceae and Stylidiaceae.

Some seeds are non-dormant. Usually, these seeds are held in the plant canopy inside protective woody fruits for many years and only released into an environment suitable for germination, often following a disturbance such as fire. As mentioned earlier in this chapter, these are known as serotinous plants and they include species of *Banksia*, *Eucalyptus*, *Hakea* and *Melaleuca*.

To germinate, non-dormant seeds need sufficient water and oxygen, appropriate temperatures and the correct light environment. The optimum temperatures for germination usually coincide with the period of reliable rainfall in the natural habitat. For species from the south-west of Western Australia, for example, germination is greatest at around 10 to 20°C, corresponding to the average temperatures during the winter wet season. Conversely, for species from northern Western Australia where rainfall occurs predominantly during the summer months, germination is greatest at between 25 and 35°C. Taking into account the type of dormancy a seed possesses, and the climatic conditions from which it originates, can greatly enhance germination success.

Chapter 7

Tissue culture and cryopreservation

Eric Bunn

Tissue culture is used all over the world for mass production of clonal plants. It involves multiplying shoots under sterile conditions in closed containers (usually glass or plastic jars or tubes) on an artificial nutrient medium. This nutrient medium contains minerals, organic compounds (sugar, vitamins, growth factors, plant growth regulators) and agar. The containers together with the medium are autoclaved ('cooked' at high temperature and pressure) to kill all micro-organisms such as bacteria and fungi which, if not eliminated, would otherwise overgrow and foul the medium.

Starting tissue cultures also requires sterilising the surface of small pieces of plants (for example, shoots from mature plants or seeds, seedlings or seed embryos) using chlorine compounds such as bleach. The sterilised plant pieces are then placed on the autoclaved nutrient medium. Sterilising the plant material is necessary to get rid of micro-organisms that normally live on the surface of the plant, which would otherwise grow out of control in the nutrient medium and kill the shoot before it had a chance to grow. Tissue-cultured plants are grown under conditions of controlled temperature (25°C) and lighting to maintain healthy growth.

Eric Bunn

Native species in tissue culture at Kings Park that are threatened with extinction in their natural environment.

After a suitable time under controlled conditions of temperature and light in a laboratory culture room, the pieces of plant material begin to grow and multiply on the nutrient medium. They are then separated into single shoots under aseptic conditions in a sterile air cabinet or laminar flow, and placed on more nutrient medium. In this way one shoot can be multiplied into many shoots with each successive subculture period (usually 4 weeks), thereby producing many thousands of identical (clonal) shoots per year. Different size tissue culture containers may be used to prepare shoots for various phases of culture, especially just prior to root induction and transfer to the soil.

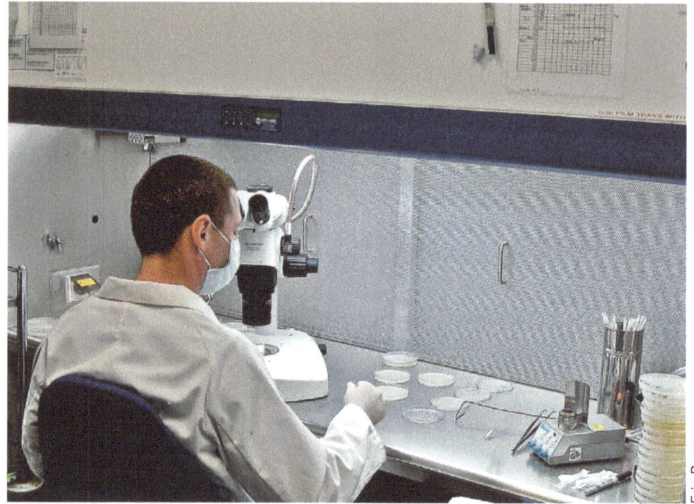
Eric Bunn

A laminar flow where operators carry out all tissue culture manipulations in a steady stream of micro-filtered sterile air.

By manipulating the compounds that regulate plant growth (with plant growth regulators, sometimes referred to as plant hormones) shoots are first multiplied using one class of plant hormone, then converted to whole-rooted plantlets (using a different class of plant hormones), which can then re-join the 'outside world'. They are transferred from the tissue culture containers into potting mix and grown in a glasshouse with controlled humidity (for example, fogging) and temperature until they fully adjust to ambient conditions outside the culture environment. The whole process of plant shoot culture and mass-producing cloned, rooted plants capable of surviving in soil is called micropropagation.

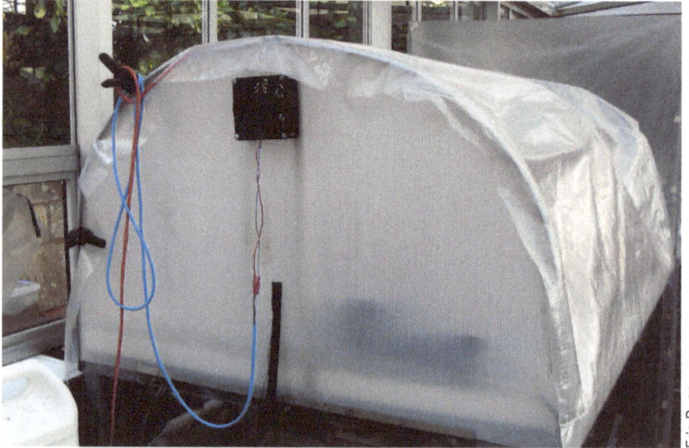
Eric Bunn

Rooted plants from tissue culture are placed in soil and kept in a fogging tent until hardy enough to survive in the nursery.

Micropropagation is the most common form of tissue culture as most plants can be grown effectively through this process. A number of native species such as kangaroo paws (*Anigozanthos* spp.) have been micropropagated successfully for many decades using the methods just described. However, some species do not perform well in standard shoot tissue culture, so micropropagation is very difficult or impossible to achieve on an economical scale. Other avenues for *in vitro* propagation must then be used, including regeneration from parts of the plant other than the shoots. One such method is somatic embryogenesis.

Somatic embryogenesis

Somatic embryogenesis refers to the development of fully formed seed-like embryos directly from somatic tissues. That is, the resulting embryos form directly from plant tissues (or cell aggregates derived from plant tissues) implanted in an agar medium, rather than from the fusion of a pollen cell with an egg cell in the ovary of a flower that produces a seed embryo. Somatic embryos have both root and shoot axes (that is, bipolar structure) as per zygotic or seed embryos; therefore the need to induce rooting as a separate phase is bypassed.

Maturing somatic embryos develop and germinate into seedlings that can be transferred to soil in the same way as zygotic seedlings. Many thousands of somatic embryos can be produced from as little as one gram of plant tissue, making this process very efficient and highly productive. However, the process requires optimising for each species and it can often take considerable research time to develop successful protocols.

Eric Bunn

Somatic embryogenesis can produce thousands of tiny somatic embryos (less than 1 mm in length) from a single gram of plant tissue.

Uses of tissue culture for plant propagation and cryopreservation

Micropropagation is used to produce plants for agriculture, forestry, horticulture, plant breeding and for conservation purposes. In plant breeding, tissue culture is used for rescuing and growing hybrid zygotic embryos, microspores (micropropagation of haploid plants), generating somatic hybrids using protoplast fusion methods, and mutation breeding using mutagenic chemicals or radiation.

Mass propagation of plants is achieved through regenerating and multiplying cultivars, hybrids and genetically modified plants by micropropagation and/or somatic embryogenesis. Disease-free or disease-resistant stock plants are produced via meristem isolation and culture (for virus-free plants); and by *in vitro* grafting (to disease-resistant seedling rootstocks).

In addition, tissue culture is used to provide *in vitro* grown shoots for cryopreservation for long-term storage of valuable cultivars of agricultural or horticultural species and for threatened plants.

Micropropagation research for conservation of threatened plants

In the case of micropropagation of endangered plants for *ex situ* (that is, 'off-site' as opposed to *in situ* or 'on-site') conservation purposes in Australia, this is almost exclusively done by conservation agencies or botanic gardens funded by state or federal governments. Kings Park is one of a number of botanic gardens around the world that has *in vitro* research capabilities for this type of work. Since the mid 1980s Kings Park has used tissue culture to conserve *ex situ* collections of many threatened plant species where seed and/or cutting material was not available. A number of these programs have led to successful restoration of threatened plants back into the natural environment, initially using micropropagated plants.

Once populations are stabilised and plants begin to produce their own seed the species are usually back on the road to recovery. If more plants are required for restocking at a later time, these can be readily sourced from tissue culture stocks or revived from cryopreserved shoot tips.

Cryopreservation research for threatened plants

Kings Park also has a cryogenic storage research capacity that complements the *in vitro* production of rare species. *In vitro* material can be safely and efficiently cryostored in liquid nitrogen for long periods of time, thereby significantly reducing the costs of maintaining *ex situ* germplasm collections of threatened plants. Plant cryopreservation encompasses the storage of specially prepared plant tissues in liquid nitrogen at −196°C so that the stored tissues retain their viability following re-warming and regrowth.

The key preparation of plant material (usually shoot tips) involves pre-treatments with chemicals that prevent freeze damage to the tissues of the shoots. In this way the shoot tips can be frozen in liquid nitrogen and later warmed to room temperature. They can then resume growth under specific *in vitro* conditions and be grown into thriving plants. The great advantage of cryopreservation is that once the shoot tips are frozen in their small cryovials, they can be stored in liquid nitrogen for very long periods (technically forever) with almost no maintenance required, with significant cost savings compared to standard tissue culture that requires monthly subculture of shoots and is prone to culture loss through operator error such as accidental contamination. Cryopreservation is the ultimate low-cost, efficient form of storage for plant material including shoot tips, small seeds and pollen, and one cryogenic dewar storage unit can store thousands of samples.

Eric Bunn

A cryo dewar that can hold thousands of plant samples for long-term efficient storage of rare and threatened plants.

Conclusion

While tissue culture is almost indispensable for plant breeding and biotechnology, micropropagation of woody plant species remains relatively expensive for horticultural purposes and the technology therefore tends to be mainly used for elite, high-value cultivars to be cost-effective and competitive in commercial markets. The production of disease-free stock plants for agriculture and horticulture is another valuable application of micropropagation that is widely utilised. Many more species can be micropropagated and make spectacular potted plant displays, but most remain to be commercially developed. Continuing research into somatic embryogenesis keeps the door open for development of industrial-scale plant production for restoration with agro-forestry applications where reducing the cost of *in vitro* plants is a primary objective.

Tissue culture has been used extensively over the past 60 years for mass propagation of many thousands of plant species, cultivars and hybrids for horticulture, agriculture, plant breeding and forestry applications. As well, tissue culture has come to play an important role in preserving rare and threatened plant species, as evidenced by the presence of tissue culture research facilities in a number of botanic gardens around the world.

Eric Bunn

Native species of potential horticultural value, such as triggerplants (*Stylidium* spp.), can be produced via micropropagation.

Tissue culture combined with cryogenic storage is an extremely useful blending of two technologies that enables ultra-long-term, safe and efficient *ex situ* storage for a large range of threatened plants, but it is also extensively used to preserve key agricultural and horticultural species, including many cultivars that would otherwise be lost over time due to 'genetic erosion' or somaclonal variation inherent in long-term repetitive vegetative propagation. This *ex situ* storage capability can also be brought to bear on the ever-increasing number of critically endangered Australian native plant species where there are no other options for propagation and storage.

There is ample scope for tissue culture and cryopreservation to improve not only conservation results but also the development of Australian native plants for sustainable horticulture. As technology improves this scope will continue to expand and deliver exciting new outcomes for propagation, long-term storage, breeding and development with Australian plants.

Kings Park continues to conduct biotechnology research into tissue culture and cryogenic storage for endangered and recalcitrant indigenous flora as part of the overall objective of conservation of plant diversity in Western Australia.

Eric Bunn

This Western Australian sedge is horticulturally attractive, drought-tolerant and disease (dieback) resistant. Somatic embryogenesis can produce such plants more cheaply than conventional shoot culture can.

Chapter 8

Pests, diseases, disorders and other problems

Aileen Reid, Bill Woods, Kevin Seaton and Elaine Davison

This chapter aims to give a general overview of problems with Australian plants, providing a summary of symptoms from pests, pathogens or disorders that affect the physiology and function of the plants. There is also more detailed information about some of the specific problems, together with suggestions of management options. The information presented here is drawn from experience, observations and research with Australian plants over several decades. Although some aspects are specific to Western Australia, there are general principles that are more widely applicable relating to the diagnosis of the cause and the management of the problem.

Know your plant and its environment

It is important to know your plant in order to recognise what is normal. Where does it grow naturally, in which climatic conditions and what type of soil? Does it usually die back after flowering or does it have a period of natural dormancy? Does it yellow in winter and recover in spring as the weather warms up? What is its usual growth rate and life span?

Some plants, such as wattles, are renowned for their short life spans, but most woody plants are long lived. During their life they will have to survive hail, storm damage, predation by insects and attack by pathogens. Their natural defences include chemicals that make them unpalatable, physical barriers such as bark that are difficult to breach, and a suite of active responses that compartmentalise damaged tissue.

Elaine Davison

An example of compartmentalised wood rot.

However, if the plant is not growing well, perhaps because of extreme weather conditions, it will not be able to respond as rapidly to attack by pests and pathogens as a vigorous plant would.

Soil factors

A plant growing in a duplex soil or in a soil with an impeding layer in the profile will usually have a shallow root system to take advantage of the seasonal, perched water-table above the impeding layer. Such a plant is susceptible to being blown over by wind during very wet weather and to drought stress in hot weather.

Root-binding

A major cause of plant failure is poor root function caused by root-binding (which is also referred to in Chapters 1 and 4). Frequently when plants are propagated they are left in their pots far too long. However, the problem starts during the first few weeks of propagation when developing roots can curl around, either seeking air if overwatered or deflected sideways due to the shape of the propagation cell, with round ones being the worst. Once root-binding starts to occur little can be done to stop the problem getting worse. Leaving this tube stock in the propagation cell for several months only magnifies the problem. Planting root-bound plants compromises their survival and eventually the tap or main root forms a knot by twisting around itself as it expands.

DAFWA

Root-binding was the cause of death for this waxflower plant.

As a result, the root system is weak at the root crown, forming an entry point for various rots and disease. After several years a mature plant can snap off at ground level, particularly in windy situations. The other consequence of root-binding is that the root system is restricted and unable to spread out and explore the surrounding soil. This compromises the whole growth of the plant as it starves of nutrients and water, which may show up in a yellowing of leaves, or eventually plant death. The best approach to this potential problem is to tap seedlings out of propagation tubes and examine the root system for twisting and curling. If this has occurred the tube stock should be discarded. Use square propagation tubes with root trainer fins and air-pruning capacity. Prepare the planting bed ahead of time and put plants into the ground as soon as possible after purchase.

Environmental factors

Extremes of weather such as exceptionally high or low temperatures are also important when considering plant health. High temperatures can lead to the rapid desiccation of leaves and twigs and it is easy to make the association of leaf damage with periods of hot weather. Frost damage may be more difficult to recognise because it can be patchy, depending on topography. Canker fungi often invade frost damaged shoots, resulting in extensive shoot dieback. Frost damage over several seasons may damage the cambium layer, resulting in plant death some time later.

When deciding whether environmental factors have contributed to observed symptoms, it is important to consider what has happened in the previous months and years, not just in the past weeks. Following severe drought, for example, eucalypts often appear to die suddenly, almost overnight, but this is the final stage of dehydration that has probably been happening for months, with the foliage surviving on water stored in the branches and trunk. It is only when this stored water has been exhausted that the foliage dies.

Symptoms such as unseasonal flowering, distorted leaves, wilting shoots or plant death indicate that something is wrong. Some symptoms such as leaf discolouration, wilting and death can have multiple causes, including poor nutrient management, herbicide damage, poor irrigation practice, poor root structure, or root damage from pests and pathogens. Detailed examination and laboratory analysis are usually required in order to make a diagnosis and suggest how the problem(s) can be minimised. This is often not straightforward, especially when more than one problem becomes apparent. Which is the primary cause and which is secondary? Is poor root structure more important than root damage from pathogens? Are fungicides the best way to manage botrytis in a nursery, or should management focus on water quality and fertiliser application to ensure that seedlings are hardened off before seasonal conditions favour botrytis? These are examples of issues that need to be teased apart in order to differentiate between primary and secondary causes.

Plant nutrition

Australian plants have widely differing nutritional requirements. Some are adapted to poor soils while others are quite tolerant of relatively high salt levels (not only sodium or chloride but also various salts that occur in fertiliser). Even within a genus there is sometimes wide variation. For example, *Verticordia plumosa* tolerates irrigation with solutions up to 5.5 dSm^{-1} while other species such as *V. nitens*, *V. eriocephala* and *V. cooloomia* experience growth suppression at levels only one-quarter of that concentration.

Fertiliser application is best spread throughout the growing season, with more added during flushes of growth.

Many species are also phosphorus insensitive and this is not necessarily restricted to the Proteaceae. For example, *Macropidia fuliginosa* is less tolerant of phosphorus than other species of kangaroo paw, and *V. plumosa* is less tolerant of phosphorus than waxflower (*Chamelaucium uncinatum*). Symptoms of phosphorus toxicity vary from blackening of leaf tips in kangaroo paws (not to be confused with alternaria infection or ink disease) to yellowing of foliage similar to iron deficiency.

For those plants that are intolerant of phosphorus, animal manures should be used with caution as they contain relatively low ratios of nitrogen to phosphorus. Both composts and manures can contain up to 40 per cent of their phosphorus in the water-soluble form, so the phosphorus content needs to be considered in the same manner as any other chemical fertiliser.

Irrigation practice

A common belief is that Australian native plants do not require irrigation. However, this depends on many factors including the provenance of the plant and how well you want it to grow and look. A boronia from the south-west forest regions of Western Australia will require much more water than a northern sandplain banksia.

Irrigation failure, particularly in drip systems, is fairly common. Many systems are installed sub-mulch or sub-soil and it may not be evident when an emitter is blocked, until a plant starts to struggle or dies. The fault may not be at the emitter; breakages or damage from rodents or nearby digging should also be considered.

In drip irrigation systems, sometimes the plant simply outgrows the system. Where one dripper was initially installed, two or three may be required as the plant matures in order to supply enough water. Plants may die when emitters are moved suddenly, especially during hot weather, since the root system will have developed where there was available water.

With rainfall becoming increasingly variable, attention to irrigation scheduling is becoming increasingly important. It is not uncommon for daily evaporation to vary tenfold within a month and plant health during a dry spell in spring should be carefully monitored. Cool temperatures can still result in relatively high levels of evaporation when the humidity is low or during windy periods.

Water stress

Many other factors apart from irrigation can result in a plant experiencing water stress. Non-wetting soils are a common cause of problems. They may occur on sands as hydrophobic organic matter builds up in mulch layers on the soil surface or around individual sand particles. Potting mixes may also become non-wetting, either in the pot before planting or in the soil afterwards. Since non-wetting can be patchy it is important to check closely around the root ball of the problem plant to exclude it as a cause.

Pesticide and herbicide use

Drift from herbicides should always be considered as a possible cause of plant symptoms such as yellowing or other abnormal colouring, growth distortion, stunting or death.

DAFWA

Symptoms of herbicide (dicamba plus MCPA) drift on to waxflower.

While spray damage from contact herbicides is more easily identified, drift may come from off-site sources that can even be several kilometres away. Some of these herbicides can cause stunting for several years.

In urban areas, neighbours spraying in windy conditions or close to fences or verge spraying can affect nearby plants. Soluble herbicides such as diuron are often sprayed around buildings but can easily affect roots of nearby trees.

Elaine Davison

Symptoms of diuron toxicity on a eucalypt leaf.

Aerial drift is not the only concern. Atrazine is often used to control weeds in paths and can affect plants in nearby garden beds as it moves through the soil.

Though glyphosate has been considered relatively non-toxic in the past, it has been shown to significantly decrease the availability of manganese, iron and zinc in soils. In sandy soils and where plant roots are close to the surface, glyphosate and many other foliar spray herbicides may adversely affect or kill plants as the herbicides encounter the roots before breaking down.

Management

Determining if a symptom is sufficiently serious to require active management depends on the context in which it is observed. Unless a problem results in death it may be tolerated in a landscape or garden planting where cosmetic damage is insignificant. However, for anyone who is growing Australian plants for their value as cut flowers or cut foliage to the floristry industry, even minor cosmetic damage can be very significant. Also, for growers wishing to export interstate within Australia or overseas, the presence of any pest or disease may contravene phytosanitary arrangements and the plant material cannot be exported.

While management of pests and diseases is discussed in this chapter, chemical controls have not been included since they frequently change. The Australian Pesticides and Veterinary Medicines Authority website (www.apvma.gov.au) provides up to date information on chemical registrations and minor use permits.

Symptoms affecting a whole plant or a large part of a plant

Death of a whole plant may be due to:

- Water stress (for example, from reticulation failure, from insufficient irrigation or from sufficient water applied to a non-wetting soil)
- Waterlogging
- Herbicide
- Frost damage
- Mechanical damage (for example, root damage from nearby excavations, whipper-snipper damage or lawnmower damage)
- Root rot caused by fungal pathogens
- Termite attack.

Death of one side of a plant may be due to:

- Root-binding
- Mechanical damage (for example, root damage from nearby excavations, whipper-snipper damage or lawnmower damage)

- Insect borers
- Root rot caused by fungal pathogens.

Poor growth may be due to:

- Poor nutrition caused by insufficient fertiliser being applied or made unavailable due to poor irrigation practices
- Water stress (for example, from reticulation failure, from insufficient irrigation or from sufficient water applied to a non-wetting soil)
- Root damage due to pests or pathogens or mechanical damage from nearby excavation
- Root rot caused by fungal pathogens
- Poor root structure due to soil compaction or impervious layers or as a result of root-binding
- Wind damage resulting from a lack of protection from prevailing winds, poor plant structure (for example, top-heavy) or planting too shallow.

Wilting may be due to:

- Water stress (for example, from reticulation failure, from insufficient irrigation or from sufficient water applied to a non-wetting soil)
- Waterlogging
- Salt stress resulting from excessive fertiliser application, poor irrigation practice or a change in the quality of irrigation water
- Root damage due to pests or pathogens or mechanical damage from nearby excavation
- Root rot caused by fungal pathogens.

Symptoms affecting individual leaves or flowers

Yellowing or reddening of leaves may be due to:

- Nutritional problems. These are not always the result of inadequate supply of nutrients; often there are other factors in the root zone impacting on the supply of nutrients such as high pH (caused by iron, manganese and zinc deficiency) or low pH (caused by magnesium and potassium deficiency).
- Waterlogging can affect not only the uptake of nutrients but can also convert them to reduced forms (for example, iron or zinc) which can be toxic.
- Many viruses and also herbicide damage can induce colour changes in leaves.
- A nearby plant may have been removed or pruned, resulting in changed light conditions.
- In eucalypts, Mundulla yellows has yellowing of foliage as one of its symptoms. This is now believed by many to result from high pH conditions in the root zone (often from limestone put down as part of road-making operations).

Distortion or physical damage of leaves and flowers may be due to:

- Insects (for example, mites, thrips, aphids, leaf-rollers, leaf-webbers and gall-forming insects)
- Windblasting or sandblasting
- Rodents climbing up onto trees and shrubs and feeding on leaves or flowers, resulting in a range of symptoms depending on the stage of the plant part at the time of attack
- Birds, especially cockatoos
- Unseasonal temperatures and often widely varying day/night temperatures
- Frost.

DAFWA

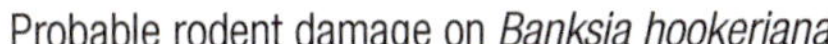

Probable rodent damage on *Banksia hookeriana*.

DAFWA

Banksia ashbyi blooms deformed by frost damage.

- Oedema can be confused with the symptoms of windblasting when the cells of the epidermis become lignified, giving a grittiness to the leaf surface. Oedema occurs when transpiration becomes impaired relative to water uptake and cell rupture occurs. A sudden bout of humid weather can provoke oedema. Plants grown under shade-cloth may also develop the condition when air circulation is impaired, allowing humidity to build at certain times.
- Infection by phytoplasmas or mycoplasmas may distort flowers and/or leaves.
- Rust galls produce characteristic growths on flowers and leaves.

Spots or blotches on leaves or flowers may be due to:

- Fungal or bacterial pathogens
- Virus
- Nutrient deficiency or toxicity
- Spray damage. Desiccant herbicides produce brown, 'burnt' lesions in a characteristic spray pattern if they drift onto a non-target plant.

Browning or scorching of leaves may be due to:

- Salt stress resulting from excessive fertiliser application
- Thrips feeding
- Water stress or a change in the quality of irrigation water.

Leaf drop may be due to:

- Transplant shock
- Frost damage
- Salt stress resulting from excessive fertiliser application, poor irrigation practice or a change in the quality of irrigation water.

Symptoms predominantly on bark or twigs

Cracks in the bark which remain clean may be due to:

- Frost damage
- Sun burn
- Lightning damage.

Cracks in the bark accompanied by gumming or oozing are generally the result of infection by fungal or bacterial canker pathogens.

Blights or dieback of twigs may be due to:

- Canker-causing fungal or bacterial pathogens
- Other pathogens such as alternaria
- Frost damage
- Wind causing desiccation and death of young soft tissue
- Drought (water stress).

Deformed twigs may be due to:

- Infection by a phytoplasma
- (In the case of galls) insects or fungal or bacterial pathogens.

Symptoms affecting flowers over most of the plant

Absence of flowers may be due to:

- The plant being juvenile (usually a seedling).
- Poor nutrition. Excess nitrogen can promote vegetative growth at the expense of flowers.
- Environmental (for example, light, temperature and day length). Erratic temperatures may prevent flowering in those plants where temperature is important for flower initiation and development. Plants with a day length requirement may not flower if grown in a location where the day length is too long or too short.
- Poor pruning practice. As a general rule, pruning should be done as soon as possible after flowering. Pruning too late may remove developing flower buds and so reduce the next season's flowering (see, for example, the advice on pruning *Banksia coccinea* in Chapter 1).

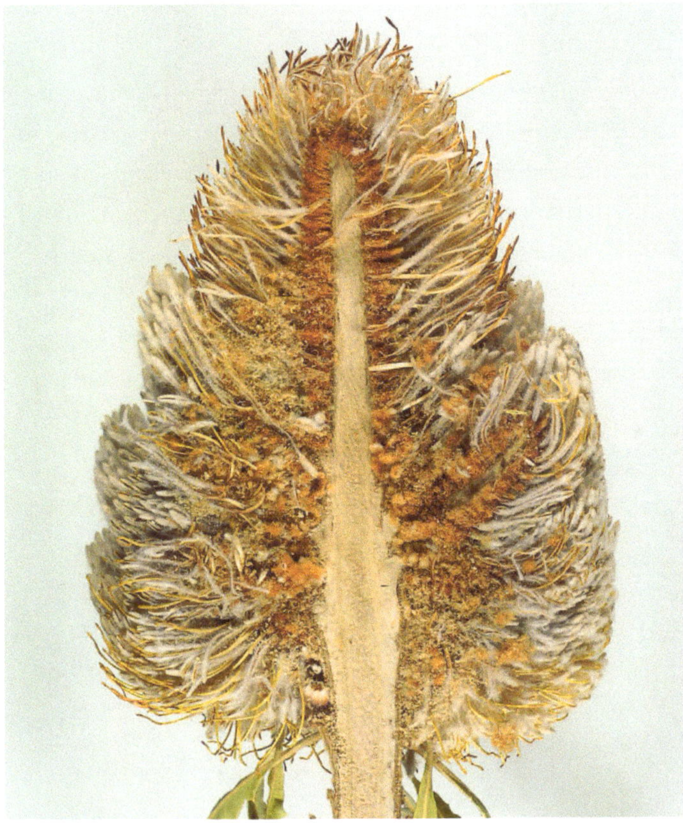

DAFWA

Abnormal flower development in banksia caused by unseasonal weather conditions.

- Unseasonal or uneven flowering is usually due to erratic weather conditions, but it can also result from untimely pruning.
- Failure of flowers to open is frequently due to blasting from wind damage, but it can also be the result of fungal infection from pathogens such as botrytis.
- Lack of petals or other malformations of flowers are frequently due to deficiencies of calcium or boron.
- Aberrations such as leaves or flowers developing in or on flowers may be due to erratic weather conditions, especially unseasonal temperatures.
- Greening of petals or of whole flowers may be due to infection by mycoplasmas.

Dropping of flowers may be natural, because some plants have an inbuilt mechanism for regulating flower load. But frequently other environmental factors are responsible, such as frost or a sudden change in salinity in the root environment due to a fertiliser application or sudden drying out of the root zone.

Diseases

Australian plants are susceptible to a wide range of diseases in the same way as exotic plants. The pathogens that cause disease may be spread as spores transferred by air, wind or water, or by insect vectors (such as aphids, thrips and leaf-hoppers) and animals. In some cases diseases are spread by mechanical contact with roots or stems. In many cases, spores are present on, or in, plants (these are called endophytes) but remain in sufficiently low numbers not to pose a threat to plant health. However, when environmental conditions are favourable to the pathogen, then spores may increase to levels that may cause disease if other conditions are met. Plants under stress from insect attack, drought, waterlogging or malnutrition are generally at increased risk of infection.

Plant pathogens vary in their ability to infect plants and some weaker pathogens may only cause disease when plants are stressed or when there is an existing point of entry through wounds or plant structures such as stomata or lenticels.

The following sections describe below-ground diseases (phytophthora, armillaria, other root and collar diseases and nematodes) and above-ground diseases (cankers, leaf-spots, rusts, mildews and moulds).

Phytophthora

Many Australian plants are susceptible to phytophthora root rot (also commonly called phytophthora dieback and dieback disease), especially members of the families Proteaceae and Ericaceae. Symptoms of infection include sudden death of plants, slow decline (often from the tips back and sometimes with only one side of the plant affected) and cankers (bark infections) at the root collar and base of the trunk. In a nursery the symptoms are root and collar necrosis. Laboratory tests are essential to confirm the presence of phytophthora.

The most widely known pathogen is *Phytophthora cinnamomi*. However, there are many phytophthoras that can infect Australian plants and cause similar symptoms. Some, such as *P. cinnamomi*, are introduced pathogens; others such as *P. arenaria* are believed to be endemic to Australia. These soil-borne pathogens are easily spread through the movement of infested soil (including gravel and sand), infected plants and irrigation water.

Phytophthoras are very important nursery pathogens that cause significant seedling losses. They are readily spread around a nursery in soil, on plants, in irrigation water, on tools and other equipment and by staff.

The best management for phytophthora is avoidance. Buying plants from accredited nurseries will avoid introducing infected plants into uninfected areas. Also, to avoid contaminating clean ground, check the source of all materials, including mulch and material such as clay and gravel for driveways and hard standing areas. Contractors such as drillers and gardeners may also bring in phytophthora in dirt on vehicles, footwear and tools. Irrigation water from sources in contact with soil, such as a dam, soak or creek, may also be a source of infection.

It is not possible to eliminate phytophthora from soil once it has been introduced. The most common method of dealing with phytophthora is to suppress the disease using phosphorous acid (phosphite) applied by trunk injection or as a foliar spray at times of active root growth. The chemical is absorbed by the plant and, in responsive plants, phytophthora invasion is rapidly contained. Not all plants respond to phosphite and some respond only to either trunk injection or foliar spray. Some plants are damaged by phosphite. Phosphite does not kill phytophthora in soil.

Armillaria

Armillaria luteobubalina (honey fungus) is a native Australian fungus that is widespread and common in native vegetation. It has a wide host range that includes introduced woody plants as well as Australian trees and shrubs. Symptoms include sudden death of small trees and shrubs, slow decline of trees and basal cankers that can be mistaken for fire scars. Armillaria mushrooms may develop at the base of infected trees during the appropriate season. These mushrooms are honey-coloured with a rough cap that feels like a cat's tongue. The gills are honey-coloured and there is a white or cream ring on the stem.

Elaine Davison

Armillaria mushrooms.

Armillaria infects roots and spreads from host to host by root contact. It grows as white mycelium in the cambium, between the bark and the wood, causing a white rot in the sapwood. The infection can spread up the trunk, killing the bark and causing a basal canker. These cankers expand if the tree is under stress from environmental conditions; however, the tree can also contain these cankers when it is growing well. Thus trees can be chronically infected for many years, with the infections waxing and waning depending on the weather.

Elaine Davison

Armillaria canker.

Management of armillaria mainly involves being aware that there may be a potential problem from chronically infected trees and shrubs in areas of native bush. Infection can spread by root-to-root contact from residual trees to newly established plantings. If infected trees are felled, as much of the stump and root system as possible needs to be removed because any armillaria infections will expand rapidly, colonising the dying stumps and roots, and will be a source of infection for new plantings.

Other root and collar diseases

There are a number of other root and collar rots caused by fungal and oomycete pathogens. Laboratory identification is essential to diagnose the cause.

Rhizoctonia is a particular problem with *Ixodia* due to the top-heavy nature of the plant. Typically, plants infected with rhizoctonia collapse and die quickly from a rot at the root collar. Such infections are associated with waterlogging, wind damage or the plant having been planted too deeply. Occasionally the fungus may enter through insect damage. The best way to minimise such problems is to reduce the likelihood of water pooling around the root collar, avoid heaping mulch around the base of the plant and provide protection from strong wind.

Pythium root rot is common in wet soil and in soils of low pH or high salt levels. Plants may wilt or the leaves turn yellow. Symptoms may come and go with changing weather conditions. Most *Pythium* species are only able to invade extensively if a plant is stressed from other causes. Pythium often occurs in summer under conditions of high soil temperature and frequent irrigation.

There are many species of pythium causing root rot, many of which are ubiquitous in soil. They are 'root nibblers' which infect root tips, so plants that are more vigorous may produce sufficient roots to keep ahead of the fungus. Likewise if the soil dries out and conditions become more favourable to the plant, new roots may be produced and the plant may temporarily recover.

To effectively control pythium, any soil problems relating to pH, salinity and waterlogging must be resolved before chemical control methods will be effective.

Plant pathogenic nematodes can also damage roots. Like fungal root pathogens these can only be identified specifically by laboratory examination. Root knot nematode caused by *Meloidogyne* species is very common and causes root galls. Other nematodes are *Pratylenchus* (root lesion nematode) and *Paratrichodorus* (stubby root nematode). Above-ground symptoms are non-specific; infected plants grow poorly and may wilt prematurely. When roots are examined, root galls (stubby roots and partly rotted roots) may be apparent. Care must be taken not to confuse root galls with rhizobium nodules on legumes and stubby branched roots with proteoid roots.

If nematode problems are suspected before planting, the soil can be treated with a fumigant. However, fumigants are highly toxic and generally not permitted for use in domestic situations.

Another method for reducing nematode numbers is by rotation with non-hosts, but this may not prove to be very effective or practical because most nematodes have very wide host ranges.

Cankers

Cankers are primarily infections of the bark, although some infections extend into the underlying wood. Infections can occur on twigs and small branches, leading to crown dieback.

They can also occur on trunks and large branches, causing large patches of dead bark. Some trunk cankers are very superficial; others may be so deep that they reach the cambium and persist for many years. Trees such as eucalypts respond to cambial damage by the production of copious amounts of gum. Excessive gumming may be a symptom of canker, but gumming may have other causes such as insect attack, mechanical damage and frost damage.

Elaine Davison

Twig canker.

Aileen Reid

Canker symptoms on eucalyptus.

Fungi are the most common cause of cankers, but bacteria can also be a cause. Canker pathogens are widespread and common in tree crowns, especially on old suppressed branches and twigs. They do little damage to a healthy tree or shrub. Canker pathogens usually fruit in the infected bark, and the fruiting bodies appear as pimples in the dead bark. There are many different pathogens that cause cankers, and these can only be differentiated by laboratory examination. Some canker pathogens such as *Botryosphaeria* have a very wide host range infecting both Australian and introduced woody plants, while others such as *Quambalaria* species and *Cryptodiaporthe melanocraspeda* are restricted to a small number of native species.

Canker pathogens cannot infect through intact bark. Infection is through wounds, such as those caused by insects, hail damage, frost cracks, mechanical damage including pruning, bark splits and at branch junctions. Infections are usually walled off quickly by a healthy plant. However, compartmentalisation may be slower or incomplete if the plant is affected by environmental stress.

There is no easy way to manage canker pathogens because they are ubiquitous. The best way is to ensure that the plant is growing vigorously, and to minimise any damage that can provide entry points.

Leaf-spots

Many different fungal pathogens cause leaf-spots. They tend to be more prevalent when the weather is humid or showery. While they can be unsightly they generally do not kill plants. Laboratory diagnosis is essential to determine the best method of control.

Some such as *Alternaria* leaf-spot and stem blight are important on kangaroo paws (ink disease) and waxflower. On kangaroo paws, alternaria causes large black spots and blotches. There is a wide range of susceptibility between species and hybrids, with *Anigozanthus flavidus* being the most susceptible. In waxflower, alternaria causes small necrotic spots on leaves and stems that are 1 to 2 millimetres in diameter with a red border. If infection is severe these spots can coalesce, ringbarking the petiole or twig and resulting in premature leaf fall.

DAFWA

Alternaria infections on waxflower.

Tar spot (*Phyllachora grevilleae* subsp. *grevilleae)* is common on *Hakea myrtoides,* causing black, shiny, domed spots 1 to 2 millimetres in diameter immersed in the leaf and usually surrounded by chlorotic haloes. The disease develops slowly over several months with infections starting in autumn and the sexual stage maturing over winter and spring. Dispersal is likely to be by rain splash, and free water is needed for ascospore germination and infection. Therefore burning the foliage may provide a means of reducing inoculum levels and providing some control. In situations where burning is impractical or prohibited, cut off all the infected foliage and remove from the site.

Vizella species are pathogens that cause leaf-spots with prominent, dark, irregular to round lesions on the foliage of grevilleas, hakeas and other members of the Proteaceae.

Aileen Reid

Vizella on *Grevillea* 'Supreme'.

Rusts

Myrtle rust, caused by the fungus *Puccinia psidii*, is a newly introduced pathogen to Australia. It is native to South America and infects many species from the Myrtaceae, including eucalypts, bottlebrush (*Callistemon* spp.) and tea tree (*Melaleuca* spp.). It forms small pustules of orange spores on infected leaves, shoots, flower buds and fruits. These spores are dispersed by the wind. However, the most common way myrtle rust has been spread in Australia is on infected, but symptomless nursery plants, and by people and machinery.

Vincent Lanoiselet, DAFWA

Myrtle rust.

One of the more distinctive rusts infecting native plant species is caused by species of the genus *Uromycladium* which infects only acacias and causes reddish-brown globular galls on young stems and shoots. The galls are woody and can reach over 500 grams, weighing down branches. Smaller galls may infect flowers, phyllodes and fruits. Often a number of insects can be found living in and on the galls. The only treatment is to prune off the affected plant parts.

Aileen Reid

Rust galls on acacia.

Another rust is boronia rust, caused by *Puccinia boroniae*, which causes rust-coloured pustules on infected leaves (which eventually turn red and drop off), and the stems and calyces of brown boronia (*Boronia megastigma*) and red boronia (*B. heterophylla*).

There are other rust species that infect grevilleas and hakeas but these are not commonly found.

Powdery mildew

Powdery mildews infect a wide range of species including verticordias and eucalypts. Most infection in eucalypts occurs under nursery conditions. The most favourable conditions for disease development are night temperatures of about 15°C and high relative humidity combined with day temperatures above 26.5°C and low relative humidity of 40 to 70 per cent. Powdery mildew spores are unique in that they do not require free water for germination; however, high humidity is necessary to start an infection and contact with a fine film of moisture promotes spore germination.

Botrytis

Botrytis, also known as grey mould, is a widespread problem both in the field and after harvest, especially on waxflower. Flowers are most susceptible and, when infected, develop tan-coloured lesions. Grey fluffy masses may also be present, especially under humid conditions. Leaves and stems may also become infected, and young shoots infected with botrytis may wither and die.

The pathogen may be present in the field but symptomless until after harvest when it produces ethylene, causing flower drop.

Other physical problems

Phytoplasmas cause 'witches' brooms' on casuarinas and wattles. They are spread by leaf-hoppers and there is no cure.

Aileen Reid

Allocasuarina infected by phytoplasma.

Mistletoes are parasitic plants which depend on the host plant for nutrition. There are 88 species of aerial mistletoe in Australia. Eucalypts and acacias are the most common hosts but casuarinas, banksias and many other native (and exotic) plants may also be parasitised by mistletoes. Mistletoe seed is spread by birds. Infected trees gradually die if the mistletoe is not removed. Some species of mistletoe mimic the host plant and are sometimes hard to distinguish except for the fact that they hang down in thicker bunches than the tree's own foliage. The only means of control is to prune out the infected area.

Mundulla yellows is a dieback disorder involving leaf chlorosis affecting many eucalypts. It was originally thought to be a new disease with the potential to devastate native vegetation, but no evidence has been able to confirm the role of any biotic agent and Mundulla yellows is believed to be lime-induced chlorosis resulting from the extensive use of crushed limestone in road-making or from the use of alkaline irrigation water.

Insects

The Western Australian insect fauna is poorly understood and many species remain unidentified. In the wild, native plants are subject to attack by a wide range of insect species which, in general, do not kill the plant. Insects may chew leaves, live in galls or be sapsuckers on the foliage.

Resource-rich flowers will host a wide range of species: wasps, thrips, bees and beetles which are attracted by pollen or nectar. Other predatory insects such as wasps and lacewings will in turn be attracted to the insects in the flowers. Most of these do not cause any damage but are of phytosanitary significance in Australian plants cultivated for the cut flower trade and destined for overseas export markets.

Some insect species such as moth and beetle larvae will attack the stem and roots of plants and this can cause plant death. Also, termites are attracted to stressed or dying plants and can kill them. In landscape planting, gardens and backyards some insect damage can be tolerated without the need for treatment. However, in a commercial planting of native plants for cut flowers any damage may be unacceptable and impact on the saleability of the product. Additionally, for interstate and overseas export, any insects on the flowers will cause rejection on phytosanitary grounds even though the insects are not causing damage.

DAFWA

Sampling for insects on waxflower in the field.

Thrips

Thrips are small insects varying in colour from pale yellow to black. Major plant pests, they can breed rapidly, can fly and are difficult to control with soft insecticides. Native species such as plague thrips (*Thrips imaginis*) often invade in huge numbers in spring if conditions are right. They prefer to feed on flowers, although some thrips species (such as *Australothrips bicolor*) can cause damage to foliage such as that of the Western Australian peppermint (*Agonis flexuosa*).

Pia Scanlon, DAFWA

Thrips damage to *Agonis flexuosa*.

Thrips scarify the surface sucking up the sap, and with severe attack the leaves look as if they have been burnt. Biological control agents such as *Orius* predatory bugs impact on numbers and treatment with an appropriate insecticide will give control.

Beetles and weevils

Large *Catasarcus* weevils will eat the foliage of many native plants and are difficult to kill with softer insecticides.

DAFWA

Catasarcus weevil.

Ringbarking weevil larvae of an as yet unidentified species attack the root crown of *Chamaelaucium*, *Verticordia* and *Thryptomene* species and have caused extensive plant deaths in cut flower plantations in Western Australia. Obvious symptoms occur only after the plant has been ringbarked. Little is known of their biology and control is by soil drench with suitable insecticides.

Ringbarking weevil.

Longicorn beetle larvae.

Larvae of cerambycid beetles such as longicorn beetle will attack the heartwood of *Xanthorrhoea* and some trees such as *Eucalyptus* species. Buprestid beetle larvae also burrow into stems and some form galls. Leaf-feeding beetles, including spring beetles and chrysomelids, occasionally occur in very large numbers and cause damage. However, populations generally disappear over time as environmental conditions change. In spring, scarab beetles and small nititulid beetle larvae and adults are often found in waxflowers feeding on nectar and pollen, but they do not cause any damage.

Flower beetle on waxflower.

Scales, lerps and mealybugs

Scale insects and lerps (psyllids) can have a hard scale or soft wax covering to protect them from natural enemies and desiccation.

DAFWA

Cottony cushion scale on eucalypt.

DAFWA

Psyllids on eucalypt leaves.

Mealybugs do not have this covering but exude a cotton wool-like wax for protection. All are sucking insects, and both introduced and native species can be found attacking Western Australian native plants. However, they are generally kept under control by native predators such as lacewings and ladybirds and by parasitic wasps. If additional measures are required petroleum oil sprays or systemic insecticides can achieve good control of scale insects if application is thorough. Mealybugs are more difficult to kill and may require the use of harder insecticides. If scale populations are large this can lead to the growth of disfiguring black sooty mould that grows on the sweet waste exuded by the scale insect after feeding on the plant sap.

Sucking bugs

Active growing tips of Australian plants can be damaged by introduced aphid species. Populations increase in size rapidly but just as rapidly collapse as predators such as ladybirds and parasitic wasps exercise control. If additional control is required systemic insecticides are effective or natural remedies such as soapy water or herbal sprays can be used. Other large bugs such as the crusader bug suck growing shoots and can cause these to wilt and die.

Gall-forming insects

Many different types of insect including beetles, flies and wasps can cause galls. The two most important to commercial growers of cut flowers are banksia leaf gall midge and Geraldton wax gall wasp.

Scarlet banksia (*Banksia coccinea*) is native to the south coast of Western Australia and is also grown as a cut flower. It is attacked by a gall midge (*Dasineura banksiae*) that causes disfiguring damage to the leaves with up to 15 galls per leaf. In the wild, this damage is cosmetic and does not harm the plant and the gall midges themselves are attacked by a suite of parasitic wasps. The hairy galls are 5 to 7 millimetres in diameter and occur on the underside of the leaves. Each contains one to five yellow gall midge larvae. There is only one generation per year, and from October to January the fragile adult midges emerge from pupae, mate and lay eggs in the succulent new growth. Eggs hatch into larvae which by feeding on the plant encourage a new gall to be formed.

DAFWA

Galls on *Banksia coccinea* caused by *Dasineura banksiae*.

Older leaves of firewood banksia (*Banksia menziesii*) also often have many distinct galls on the underside of the leaves, causing them to deform. These are caused by as yet unidentified eriophyid mites that attack the leaves when they are young and soft. As the leaves grow the galls also increase in size and remain once the leaves are hard and mature.

Aileen Reid

Eriophyid mite galls on *Banksia menziesii*.

The Geraldton wax gall wasp (*Oncastichus goughi*) is a tiny black wasp with red eyes which lays its eggs into the young tips and stems of Geraldton wax (*Chamelaucium uncinatum*). When the larva emerges from the egg and starts to feed, the plant reacts by accelerating growth around the larva and a gall is the result. Once the larva reaches full size it pupates and an adult wasp emerges from a small hole chewed in the gall. As there are several parasitic wasps that attack larvae within the galls, the wasp that emerges from a gall may not be the gall wasp itself.

DAFWA

Galls caused by *Oncastichus goughi* on waxflower.

This gall wasp is native to Western Australia and is not known to attack any other plants in the wild other than Geraldton wax. In cultivation it attacks all Geraldton wax types but seems most abundant on white waxes such as 'Alba'. Some waxflower hybrids have been attacked although there is a high degree of resistance in some crosses, particularly *C. floriferum* hybrids such as 'Lady Stephanie' and *C. megapetalum* hybrids such as 'Esperance Pearl'.

Galls vary in size from large (15 millimetres in length) to small (2 millimetres in length) and appear on stems or leaves. Stem galls can cause stems to bend, giving an obvious deformed branching pattern. Gall wasps have spread to waxflower growing areas around the world.

Moths

Larvae of moths will attack the foliage of many Australian plants, causing minor damage. An as yet unidentified tortricid moth webs together the leaves of waxflowers.

DAFWA

Tortricid moth webbing on waxflower.

Other caterpillars form unsightly nests or bags which provide protection from predators. Larvae emerge from these nests at night to feed. Case moths, which make shelters out of twigs, are often seen on native plants but cause little damage.

DAFWA

A case moth shelter.

Cup moth larvae have urticating (skin-irritating) hairs and construct protective cups from which they emerge to feed. Woolly bear caterpillars can attack plants such as kangaroo paw in winter. They also have urticating hairs for protection against predators. The banksia-boring moth (*Arotrophora arcuatalis*) is an important pest of banksias grown in cultivation. It bores into and feeds within both the developing cones and young stems. Control can only be achieved with regular application of insecticides.

Banksia boring moth damage on banksia blooms.

Larvae of large moths such as cossids feed within the stems of acacias and can cause plant death. Once inside the plant control is difficult as systemic insecticides are not very effective against large moth larvae. It can take more than a year for the larvae to complete development and emerge as adults.

Termites

In the south-west of Western Australia, termites can attack and kill native trees and shrubs. However, the plants which are attacked have normally been previously weakened by some form of stress such as drought, waterlogging or disease. In the north of Western Australia, the giant termite *Mastotermes darwiniensis* can attack and kill healthy shrubs and trees, and insecticide application is required for control.

Grasshoppers

Wingless grasshoppers are regular pests of trees and shrubs in the south-west of Western Australia. Baiting and insecticide spraying on adjacent pasture can give a degree of control. Once swarms are flying there is little that can be done to protect plants from damage. Although plants can be stripped of leaves, in many instances they will recover over time after the swarm moves on. However, in cut flower plantations even small amounts of cosmetic damage can reduce the value of a crop.

Chapter 9

The selection and breeding of Australian plants

Digby Growns and Mark Webb

The Australian native flora is generally horticulturally unimproved and many highly ornamental species are often difficult to grow due to their specialised growing requirements. Plant selection and breeding provides the opportunity to improve the adaptability of this flora under varying environmental conditions so they are more suitable for display in a wider range of public landscapes and home gardens.

The Western Australian flora contains about 12 000 species. A number of these are floristically spectacular and grow in a diverse range of climates from cool temperate to hot arid. Many species vary in form, flower colour and flowering time due to thousands of years of genetic isolation between the different populations. For example, Geraldton wax (*Chamelaucium uncinatum*), which is a very popular ornamental plant, grows over a distance of 500 kilometres in both alkaline and acid soils, with variation in form from single-stemmed plants 5 metres high to compact, multi-stemmed shrubs only 30 centimetres high, and flowering from June to December with flower colours ranging from white to purple.

This range in habitats and specialised environments often leads to significant intraspecific variation between populations of Western Australian native plant species that can be utilised for horticulture. The variation is often more evident in environments adjacent to the ocean where salt-laden and strong winds drive population selection for dwarf, compact and prostrate forms that otherwise grow upright and taller in less exposed locations.

Examples of hardy varieties from a range of environments that are adaptable to cultivation include prostrate or compact forms of *Banksia ashbyi*, *B. media*, *Boronia crenulata*, *Calytrix tetragona*, *Chamelaucium uncinatum*, *Darwinia citriodora*, *Diplolaena ferruginea*, *Grevillea pinaster*, *G. preissii*, *Melaleuca huegelii*, *M. lanceolata*, *M. pentagona* and *Pimelea ferruginea*.

This variability is not only present in plant form and flower colour, but also in hidden characteristics such as disease and pest resistance. For example, *Grevillea preissii*, a low-growing highly ornamental shrub that is popular in home gardens and landscaping in Western Australia, is normally killed by *Phytophthora* species. However, natural populations of this

Digby Growns

A dunal population of a compact form of *Chamelaucium uncinatum*.

Digby Growns

Compact forms of *Melaleuca pentagona* and *Banksia media*, Point Ann, Western Australia.

Trish Esslemont

A collecting expedition near Gracetown, Western Australia.

species have been located that are highly tolerant of the disease. Such genetic variability in natural populations can be used to provide improved disease tolerance and hardiness in hybrid progeny.

Some limited plant selection and breeding was done at Kings Park and Botanic Garden in the 1970s, but it has only been since the late 1990s that a more focused and integrated program has been established at the park. The aim of the new program is to develop plant varieties that have a range of desirable horticultural characteristics, have lower water use requirements, and can be grown in Kings Park and used more widely in home gardens and public landscapes locally, nationally and internationally.

Plant breeding and selection in Kings Park

Some past and recent Kings Park plant releases include:

- *Callistemon* 'Kings Park Special'
- *Pimelea ferruginea* 'Magenta Mist'
- *Lechenaultia* 'Lola'
- *Anigozanthos rufus* 'Kings Park Federation Flame'
- *Scaevola aemula* 'Blue Print'

The plant improvement system at Kings Park mainly involves the selection of superior plant varieties from wild populations and display plants in the Botanic Garden, and the controlled crossing of selected parents.

Plant selection

Selecting plants with superior characteristics is the most basic form of plant improvement, allowing nature to provide varietal diversity from which superior forms with desirable characteristics are chosen and tested. Under certain conditions prescribed by regulation, these superior forms can be released for commercial production and provide the basis of a targeted breeding program.

To ensure that the widest possible pool of material is used in the Kings Park program, regular field trips are undertaken to collect new material displaying the widest range of targeted characteristics. This material is then grown and tested under local conditions. Final selections are made and assessed for potential weediness. If determined to be ornamentally interesting and providing a benefit over other varieties that may already be in the market, the selected material is then tested in other environments in Australia and overseas in anticipation of a public release.

David Blumer

Callistemon 'Kings Park Special'.

One of the best known plant varieties developed at Kings Park is *Callistemon* 'Kings Park Special'. This plant was selected from a number grown from seed by the former park horticulturist and superintendent Ernst Wittwer. It remains known in the industry today as an outstanding cultivar of *Callistemon*.

Anigozanthos rufus 'Kings Park Federation Flame' was also selected from seed germinated for display at Kings Park. From many hundreds of seedlings grown, a single plant was identified that had an intense flame-coloured flower, markedly different from the red flower typical of the species. Following extensive trialling to ensure its stability and uniformity, Plant Breeders Rights were obtained and this selection is now widely grown.

A range of other selections from a number of genera including *Pimelea*, *Alyogyne*, *Hypocalymma*, *Chamelaucium* and others are continually being evaluated in the park and overseas, and it is likely some of these will eventually be grown in a wide variety of public landscapes and home gardens. As part of the evaluation process, any plant selected by Kings Park for release is assessed for its potential for weediness, to ensure there is no reasonable likelihood of it becoming a weed in another location.

Digby Growns

Anigozanthos rufus 'Kings Park Federation Flame'.

Controlled crossing

In contrast to the large number of published papers on crossing systems and processes for exotic plants such as roses, carnations and petunias, there is almost no published literature on the same practices for Australian native plants. In addition, experience at Kings Park has found multi-faceted barriers in several genera that must be overcome before hybridisation can be successful. Currently the program at Kings Park uses traditional methods of controlled crossing to develop new varieties. Under controlled crossing, pollen is collected from one parent and transferred to the stigma of another parent plant. Any fruits that develop are bagged and the seed collected for germination and introduction into a screening program. Only a very small number of the resultant hybrids make it through the screening program into advanced testing.

Controlled interspecific and intraspecific pollination builds on the natural pollination process by influencing outcomes using parent plants that are spatially separated or that don't flower at the same time. Techniques such as manipulating light and temperature to achieve out-of-season flowering and storing pollen at 2°C (short term) or at –20°C (medium term) are used to fast-track the crossing program, or to produce hybrids from parents that would not normally cross naturally.

A range of advanced seed propagation and tissue culture techniques is also used in the program to assist in the germination and propagation of hybrid material. Specialised methods in tissue culture such as embryo rescue, complete removal of the seed coat and/or use of mechanisms such as filter bridges (filter paper partially immersed in liquid media) can enhance the likelihood of successful germination. Similarly under nursery conditions, techniques such as heat, smoke, scarification and vernalisation (treatment with a period of low temperature) can be used to promote germination.

Parent plants with particular attributes such as good flower colour, early flowering or disease tolerance have been crossed with other parent plants with complementary attributes such as flowering period, flower size or plant form. The progeny of such crosses have a range of desirable attributes from which the best plants are selected for testing or further crossing.

Record-keeping is critical in any breeding program to ensure the parent plants are tracked through each phase of the hybridisation process. Many attributes are not expressed until the second, third or even later generations. For example, many white flowers are the result of a break in the metabolic pathway that normally produces coloured flowers. If this white-flowered plant is hybridised with a coloured-flowered relative, the resultant seedlings will all normally have coloured flowers. The white flower will only be re-expressed in the second generation after the first generation plants are hybridised together or back-crossed to the original white parent.

Another advanced technique used at Kings Park is genetic mapping where the relationship between one or more related genera is established through DNA analysis. This enables more closely related species to be targeted in a crossing program. Further to this, where putative hybrids are difficult to determine morphologically in the early stages of seedling growth, DNA profiles of the two parents can be compared to that of the progeny to clarify whether they are true hybrids or selfs of the seed parent. This can save significant resources that would otherwise be used to maintain plants that have little value in the breeding program.

Featured plants from the Kings Park program

Anigozanthos

Anigozanthos (kangaroo paw) is a Western Australian endemic genus with 11 species, all pollinated by birds and found only in the south-west of the state.

Kangaroo paws were among the first Australian native plants to be hybridised, and targeted programs over the past 30 years have produced a large range of hybrids popular in many countries. It is now rare for straight species to be grown, with the possible exception of *A. manglesii* and *A. viridis*. However, the genetic range available to most kangaroo paw breeding programs has been limited, and the Kings Park program is focusing on broadening the genetic base to produce hybrids with improved vigour and flowering period, disease tolerance and different colour forms.

A. flavidus is the most widely cultivated kangaroo paw, normally with yellow and green flowers on multi-branched stems or 'scapes' up to 3 metres in length. Other colour forms available include shades of red, orange and pink. *A. flavidus* is easy to grow and tolerant of a range of climates, including those with humid and wet summers. Because of its hardiness and vigour *A. flavidus* has formed the basis of most breeding programs, crossed with other more colourful kangaroo paw species that are generally more difficult to maintain in cultivation.

Selection for disease tolerance and compact forms has shown it is likely to require several generations to achieve more desirable outcomes, as the species such as *A. humilis* or *A. gabrielae* that have the best compact forms are also the most susceptible to diseases such as ink spot and rust. An example of crossing for disease resistance is shown by hybridising a selection of *A. manglesii* with ink spot tolerance with a selection of rust-tolerant *A. viridis* 'Green Dragon'. This cross has resulted in a range of hybrids with varying tolerance for each disease. Those hybrids showing the greatest disease tolerance are then selected for further breeding with the end result being a hybrid or hybrids that have multi-gene expression of the characteristic being sought.

Digby Growns

A brightly coloured variant of *Anigozanthos manglesii*. The normal red and green form is the floral emblem of Western Australia.

Digby Growns

Anigozanthos manglesii × *Anigozanthos viridis* 'Green Dragon' progeny with different disease tolerance.

Kangaroo paws are relatively easy to hybridise and depending on the parent species, can set more than 40 seeds per flower head. The pollen is easily accessible and presented at the end of the flower on the anthers, close to the stigma. In some species such as *A. gabrielae*, care must be taken when pollinating that self pollen is not transferred to the same flower. This risk can be overcome by removing the anthers of a flower prior to its pollen being released.

Digby Growns

An *Anigozanthos viridis* hybrid presenting pollen.

Kangaroo paws are divided into two subgeneric groups: those with branched flower scapes such as *A. flavidus* and *A. rufus*, and those with unbranched flower scapes such as *A. manglesii*, *A. humilis* and *A. bicolor*. Generally first generation hybrids between the two subgeneric groups have very low fertility or are sterile. This is a challenge for a breeding program, as it is the hybrids between the more disease-tolerant species in the branched subgenus and the compact species in the unbranched group that provide the best opportunity for producing a range of attractive and free-flowering forms with disease tolerance.

A range of techniques such as back-crossing, selfing and the production of tetraploid lines is used to overcome first generation hybrid sterility, as it is the second and subsequent hybrid generations (F_2, F_3, etc.) that more widely express the recombination of the range of attributes from each of the parents.

Hybrid kangaroo paw seed germinates well in response to a range of pre-treatments including heat, smoke, scarification and vernalisation. The first flowering in many kangaroo paw species and hybrids occurs as a response to germination time and plant size, independent of season. Such flowering can provide good opportunities for fast-tracking second and subsequent hybrid generations and hybridising between species that normally flower at different times.

Lechenaultia

An extensive selection and breeding activity within the family Goodeniaceae has been underway at Kings Park since the mid 1990s, firstly with *Lechenaultia* and then with *Scaevola*.

With more than 30 species, the genus *Lechenaultia* mostly occurs in the south-west of Western Australia, with a few species extending into arid and tropical climates and one (*L. filiformis*) extending into New Guinea. They are mostly small shrubs or herbaceous perennials that grow less than a metre tall with a protandrous (male stage fertile before the female stage) pollination system. This genus is celebrated for having species with outstanding, vibrant flowers across the colour spectrum.

Digby Growns

The Eneabba form of *Lechenaultia biloba*, which features darker blue flowers than more southern forms.

Lechenaultia formosa.

Many of the most spectacular species of *Lechenaultia* require specialist care and conditions to grow well under cultivation. The Kings Park program aims to develop more robust varieties with outstanding flower colours. First generation hybrids between many species are fertile, with the exception of hybrids with *L. biloba*. Hybrid seed is germinated in tissue culture using a filter bridge with liquid media, as high rates of germination cannot be guaranteed under nursery conditions.

Some of the Kings Park hybrids are suited to pot cultivation or growing as an annual or biennial plant. One of the hardiest hybrids is *Lechenaultia* 'Lola', which is a hybrid between *L. formosa* 'Champagne' and the scarlet flowered *L. laricina*. It has bright, deep pink terminal flowers and an upright form. It is suited to pot cultivation and, together with other superior hybrid selections of *Lechenaultia*, is sold by the Friends of Kings Park through their regular plant sales.

Lechenaultia formosa.

Scaevola

Scaevola is a genus of approximately 130 species, mostly found in Australia, with about 35 occurring outside Australia including species found in New Caledonia and Hawaii. They are mostly herbaceous shrubs although some species can grow larger, such as *S. spinescens* which is a woody shrub to 2 metres. Flower colour is generally purple, mauve or white, sometimes with pink variants. A few species have yellow or orange flowers.

S. aemula from Middle Island in the Recherche Archipelago. It was first collected in 1802 and forms part of the type for the species. It was rediscovered in 2007 during a Kings Park expedition to the island.

Like *Lechenaultia*, the pollination system for *Scaevola* is protandrous. In addition *Scaevola* can exhibit multi-faceted barriers to self-pollination which has implications for breeding as the genetic barriers can prevent crossing between closely related plants, particularly when using techniques such as sibling crossing and back-crossing.

The Kings Park breeding strategy is focused primarily on *S. aemula*, due to the global commercial success of *S. aemula* 'Purple Fanfare' and other related varieties.

Many hundreds of intraspecific *S. aemula* hybrids have been produced and commercially assessed. Although some interspecific progeny were bred, incompatibility issues within the genus have limited the number of wide crosses achieved.

Scaevola aemula 'Blue Print' is an intraspecific hybrid between three different cultivars and was the first *Scaevola* hybrid released from the Kings Park program. It was selected for its balance of flower and leaf; its well-branched, mounding habit that spreads as it matures; and its massed lavender-coloured flowers that give a bright, eye-catching show in a pot or in the ground from spring through to the end of summer.

Courtesy of Ball Horticultural Company

Scaevola aemula 'Blue Print', bred by Kings Park.

Grevillea

Grevillea is probably the most popular Australian genus in ornamental horticulture. It contains about 340 species, mostly within Australia but with outlying species in New Caledonia, New Guinea and Indonesia. About two-thirds of *Grevillea* species occur naturally in Western Australia.

Flower colour, leaf morphology, plant form, flowering time and geographical location vary greatly within this genus, providing a range of features to target within a breeding and development program. There have been a number of *Grevillea* hybrids released into the market over the past 15 years, underscoring their popularity. Most of these hybrids have arisen by chance in home gardens; however, an increasing number have been produced through controlled pollination.

Most of the hybrids produced are variations on the popular 'Queensland tropicals', so-called because of their geographical origin. Very little hybridisation has occurred using Western Australian species, with the exception of *G. bipinnatifida* which is one parent of the very popular *G.* 'Robyn Gordon', the other parent being *G. banksii*, a subtropical species from Queensland.

Two subgeneric groups in *Grevillea* (groups 14 and 35) have been targeted in the Kings Park program due to their range of desirable characteristics, hardiness and commercial popularity. Group 35 is the largest subgeneric group in the genus, with 61 species. It exists over the greatest geographical and climate range and has significant diversity of flower colour, inflorescence size and presentation, flowering season and plant and leaf morphology. Most of the commercially successful *Grevillea* cultivars and hybrids come from this group, including a range of *G. bipinnatifida* × *G. banksii* hybrids such as 'Robyn Gordon', 'Superb', 'Coconut Ice', 'Ned Kelly', 'Peaches and Cream', 'Autumn Waterfall' and 'Molly'. The popular Queensland tropical hybrids, such as *G. banksii* × *G. sessilis* 'Misty Pink', *G. banksii* × *G. hodgei* 'Caloundra Gem', *G. pteridifolia* × *G. banksii* 'Honey Gem', and *G. sessilis* × *G. pteridifolia* 'Sandra Gordon' are also from this group.

Digby Growns

Grevillea concinna, a group 35 species from the south-west of Western Australia.

Grevillea have hermaphroditic flowers that are protandrous. They are usually borne in groups of paired flowers forming inflorescences of various sizes. The pollen is deposited on a modified style end, known as the pollen presenter, prior to anthesis and is then accessed by pollinators once the style has elongated. Natural pollination is effected by birds or insects or, rarely, mammals. The stigma is located within the area of the pollen presenter and becomes receptive within hours or days following anthesis. Following pollination and successful fertilisation a maximum of two seeds per flower develop. Seeds germinate in response to a range of pre-treatments including heat, smoke, scarification and vernalisation.

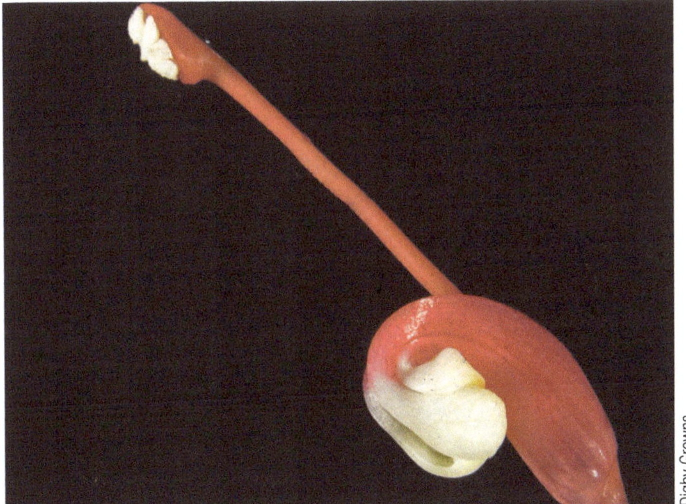

Digby Growns

Pollen presentation on *Grevillea georgiana*.

The subgeneric group 14 is endemic to Western Australia and includes well known cultivated species such as *G. olivacea*, *G. obtusifolia* and *G. thelemanniana*. These species are hardy in cultivation in Western Australia and are tolerant of phosphorus and *Phytophthora* species. Other well known species in the group, such as *G. preissii* and *G. pinaster* have variants that are hardy in cultivation. Flower colour is red in all species with the exception of *G. olivacea*, which has yellow and orange forms, and *G. stenomera* which has orange flowers.

Digby Growns

Grevillea preissii, a group 14 species.

Boronia

Boronia belongs to the genus Rutaceae of which *Citrus* is the best known member. *Boronia* is renowned for its spectacular floral displays, perfumed flowers and aromatic foliage.

The 100 or so species within *Boronia* are found all over Australia, with one species located in New Caledonia. Western Australia has about 60 species growing in seasonally wet south-western forests, the semi-arid zones of the Murchison region further north, and the northern-most tropical Kimberley region.

Many named selections are available commercially. For example, *B. heterophylla* has several named cultivars including the early-flowering 'Stella' and 'Helena Bells', the white-flowered 'Moonglow' and the claret-coloured 'Purple Rain'. *B. megastigma* cultivars include the yellow-flowered 'Lutea', the reddish 'Jack McGuire's Red', the compact 'Heaven Scent', and 'Harlequin', which has striped flowers.

Some hybrids are also in cultivation. One well known hybrid produced through an earlier breeding program at the University of Western Australia is 'Purple Jared', which is a hybrid between the sweet-scented *B. megastigma* and the floriferous *B. heterophylla*. Other hybrids in cultivation include 'Lipstick' and 'Morande Candy'; however, these are mostly chance hybrids selected from wild populations.

With nearly every flower colour represented, including the rare true blue, plus world-famous floral scents, the genus *Boronia* is an excellent candidate for a breeding and development activity. There is an opportunity through a controlled breeding program to target the more hardy species and hybridise them with those that have dense floral displays or beautiful scents. This should deliver plant varieties that are more water-efficient, combined with bright colours and wonderful fragrances.

Boronia are pollinated by insects, with *B. megastigma* the best studied species. This *Boronia* is pollinated by a moth that oviposits in the ovary of the flower, pollinating as it does so. This species, and others such as *B. heterophylla* and *B. purdieana,* have flower structures that prevent self-pollination as the anthers sit well below the stigma. Other species such as *B. crenulata* and *B. ramosa* have structures where the anthers are closely adjacent to the stigma. *B. ramosa* is a prolific self-seeder and great care needs to be taken when hybridising with this species as the seed parent.

Digby Growns

Boronia heterophylla 'Stella'.

Digby Growns

Flower structure of *Boronia purdieana* showing the pollen-bearing anthers.

Digby Growns

Insect visiting the flower of *Boronia crenulata.*

B. crenulata was the base species selected at Kings Park to produce hardy hybrids, crossing with free-flowering selections such as *B. heterophylla* and *B. megastigma*. However, no viable seed has been produced using *B. crenulata* as a parent, even at the intraspecific level either open-pollinated or through controlled crossing. This suggests this species is likely to have one or more pollination barriers that are the subject of ongoing investigation.

Viable seed has been obtained by crossing some of the free-flowering species. Hybrid *Boronia* seed is germinated in tissue culture because a nursery system has not yet been developed where high levels of germination can be guaranteed. The seed coat on hybrid seed is completely removed after initiation into culture, resulting in very high rates of germination.

Several intraspecific *B. heterophylla* and *B. megastigma* hybrids, and back-crosses to a *B. heterophylla* × *B. megastigma* hybrid are being tested. Other ornamental species such as *B. capitata*, *B. pulchella*, *B. purdieana* and the blue-flowered *B. ramosa* have been introduced into the program to broaden the genetic base.

Chamelaucium

Chamelaucium is commonly known as waxflower. It is endemic to Western Australia and comprises 31 species. *C. uncinatum*, commonly known as Geraldton wax, is the best-known species, grown in many countries as a garden and landscaping plant and valued for its cut flowers. It is a widely variable species, growing as a single-trunked plant to 5 metres, through to compact branching forms to 30 centimetres, with flowering in different populations ranging from June through to December.

Digby Growns

A multi-petalled form of *Chamelaucium uncinatum*.

C. megalopetalum (large waxflower) is less variable and more difficult to cultivate on its own roots. Kings Park routinely grafts selections of this species onto *C. uncinatum* to improve its vigour and life expectancy in cultivation. Flower size, terminal flowering, extended vase life and red flowers in some forms are the ornamental attributes that provide breeding opportunities with this species. Hybrids between *C. uncinatum* and *C. megalopetalum* are becoming more popular due to their flowering display and extended flowering.

Max Crowhurst

A selection of *Chamelaucium megalopetalum* where the flowers turn red with age.

Digby Growns

A hybrid between *Chamelaucium uncinatum* and *C. megalopetalum*.

C. floriferum has variation in form and some variation in flower colour. It is a pine-like shrub from 1 to 3 metres, mostly occurring on or near granite outcrops close to the south coast of Western Australia. It is a hardy species and there are good opportunities for selections with upright forms to be used for formal or semi-formal landscapes. Hybrids between *C. uncinatum* and *C. floriferum* are renowned for their hardiness.

Digby Growns

An upright form of *Chamelaucium floriferum* featuring pink flowers.

C. ciliatum often grows in association with granite and is a hardy species with fragrant foliage. Some forms flower as early as January with others flowering through autumn and winter, while the 'Stirling' form flowers through to late spring. Naturally occurring populations near Ravensthorpe in the south-west of Western Australia contain plants where the white or light pink flowers turn deep pink to red with age.

Digby Growns

The Ravensthorpe form of *Chamelaucium ciliatum*.

Other species in the genus with ornamental qualities include *C. lullfitzii*, *C. drummondii*, *C. axillare* and *C. croxfordii*.

As with *Grevillea*, the current strategy for Kings Park's *Chamelaucium* breeding activity is to produce hardy, easy-to-grow, free-flowering plants in a range of colours suitable for home gardens and public landscaping. Plants of *C. uncinatum* from the 'Seabird' population, which grow from prostrate plants to 60 centimetres, are the base from which hybrid varieties are being developed, using early flowering forms of *C. uncinatum*, *C. megalopetalum* and *C. floriferum*, some of which have ornamental buds while others are *Phytophthora*-tolerant.

Chamelaucium flowers deposit their pollen onto the style just below the stigma prior to the flower opening. As the style extends it presents the pollen to its natural pollinators – a wide variety of insects. In *C. uncinatum* this can result in high levels of self-pollination and the flowers therefore need to be emasculated prior to anthesis to ensure any embryos produced are of hybrid origin. However, *C. megalopetalum* does not appear to self-pollinate even when pollen is deliberately placed on thc stigma. The self-incompatibility mechanism is not currently known.

Digby Growns

Chamelaucium megalopetalum flower structure showing the pollen presented on the stigmatic hairs just below the stigma.

Max Crowhurst

Jewel beetles visiting *Chamelaucium uncinatum* in Nambung National Park, Western Australia.

All hybrid seed of *Chamelaucium* is germinated in culture as otherwise the seed remains encapsulated in woody fruits once mature, often taking years to germinate. While techniques such as scarification and applying hormones and/or smoke can stimulate germination in nursery conditions, these methods are not sufficiently reliable to be used on valuable hybrid seed. Seed is normally harvested 6 weeks after fertilisation and excised embryos are then initiated into culture. Once they germinate and grow to a sufficient size they are subcultured as micro-cuttings and eventually deflasked for growing on and assessment. First generation intraspecific and interspecific hybrids are assessed for form and flowering, and for use in further hybridisation.

Hybrids are also known between *Chamelaucium* and the related genus *Verticordia*. The intergeneric hybrids 'Paddy's Pink', 'Jasper', 'Eric John' and 'Southern Stars' are all hybrids between *C. uncinatum* and *V. plumosa*. *Chamelaucium* and *Verticordia* are part of a broader group known as the *Chamelaucium* alliance, which also includes the genera *Darwinia*, *Pileanthus* and *Actinodium*. The alliance has an amazing array of flower colour, plant form and flowering season, with species in this group found in nearly every environmental niche in Western Australia. Such diversity in the *Chamelaucium* alliance offers exciting breeding and development opportunities.

Although intergeneric hybrids within the *Chamelaucium* alliance can be produced with standard controlled crossing techniques, the level of success is extremely low. The use of somatic fusion to generate intergeneric hybrids is being investigated at Kings Park.

Somatic hybrids originate from the fusion of protoplasts. A protoplast is the living material of a plant cell after the cell wall

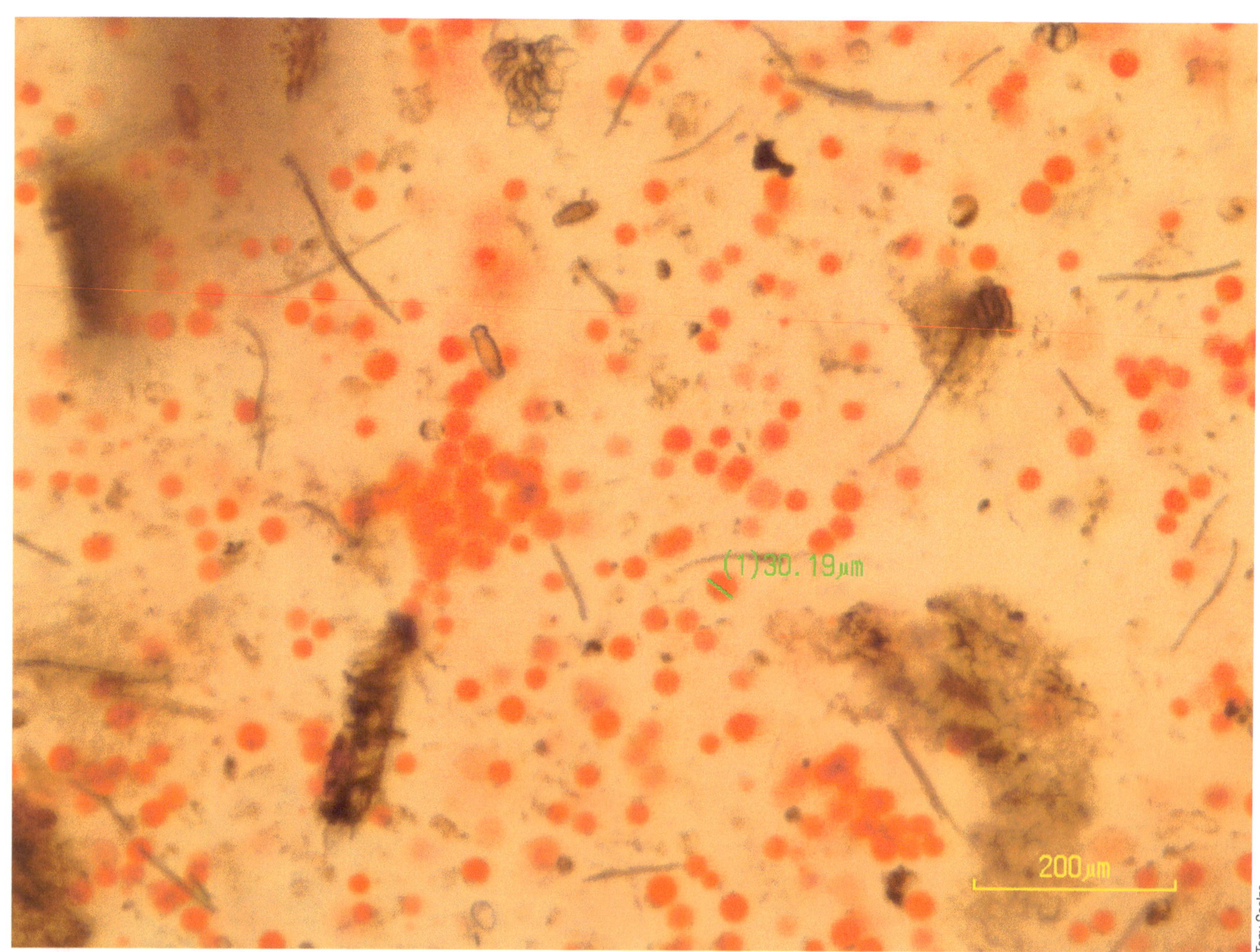

Protoplast of *Pileanthus* aff. *rubronitidus*.

has been removed. Protoplast (somatic) fusion technologies have been used in a range of breeding programs including potatoes, brassica, rice, rapeseed, tomato and citrus.

Protoplasts are combined with those of another plant to produce hybrid cells. This can be achieved either chemically by using polyethylene glycol or through electrofusion using low voltage AC and DC currents. These fused hybrids are then regenerated into whole plants through microcallus and macrocallus phases, which subsequently have hormones applied for adventitious shoot induction. These shoots are subcultured onto appropriate media for root initiation and whole plants are generated for deflasking.

The future

Over the next few years an increasing number of improved plant varieties from the Kings Park program will become available. These varieties will include superior forms selected from wild populations or from seed-grown plants, but will increasingly be hybrids developed from targeted crossing activities.

These new varieties will be tailored to cultivation in public landscapes and especially home gardens that have limited space. They will have superior flowering displays in a range of colours and will be selected for their hardiness, low water use and climate adaptability.

Glossary

ascospore a spore contained in a sac called an ascus or that was produced inside an ascus; they are only produced by fungi known as ascomycetes

arborist a person with technical knowledge and experience within the field of arboriculture. The practice of arboriculture includes the cultivation, management and study of individual trees, shrubs, vines and other perennial woody plants.

bradysporous see 'serotinous'

chlorosis yellowing of leaf tissue from the insufficient production of chlorophyll

chlorotic haloes small yellow circles on a leaf caused by infection from a pathogenic fungus

cryodewar an insulated container for liquid nitrogen

cryovial a small (usually 1 to 2 millilitres) plastic container with screw-top lid used for storing plant material in liquid nitrogen

dehisce the bursting open of certain plant structures, such as anthers, spore capsules and fruits, to release their contents

endophyte an organism, usually a fungus or bacteria, that lives inside a plant; often the relationship may be mutually beneficial but under certain conditions, it can change to one where the organism causes disease in the plant

epicormic shoots arising from latent buds or adventitious buds concealed beneath the bark of some trees (also known as water sprouts or suckers); production can be triggered by fire, pruning, wounding or root damage but may also be as a result of stress or decline

foliar of or relating to a leaf or leaves

geosporous plants that release their annual production of seeds into the soil seed bank

germplasm seeds, spores or meristematic tissue capable of repropagation usually following some form of storage (e.g. seedbanking or cryostorage), the genetic material bearing the hereditary characteristics of an organism

glaucous leaves that are covered with a grayish, bluish or whitish waxy coating that can be rubbed off, especially relates to eucalypts

globose nearly spherical

haploid one complete set of chromosomes; the condition where the chromosomes are not duplicated

inflorescence a group or cluster of flowers

meristem undifferentiated plant tissue where new cells are produced, usually at shoot and root tips

oomycetes fungus-like microorganisms also known as water moulds

operculum the bud cap covering the flowers of *Eucalyptus* and *Corymbia* trees, which falls off as the flower opens

peduncle the stem or stalk supporting a flower or inflorescence

phytoplasma	specialised bacteria that are obligate parasites of plant phloem tissue and transmitting insects (vectors)
protoplast	a plant cell with the cell wall removed
senescing	relating to plant material that has reached late maturity or is old
serotinous	plants that retain mature seeds within protective fruits held in the plant canopy, delaying seed release; also termed 'bradysporous'
somaclonal	somaclonal variation is the change in the original plant characteristics through tissue culture
somatic	non-sexual cell or tissue
zygotic	relating to a zygote; the initial cell formed when a new organism is produced by fertilisation

Further reading

Introduction

Byrne J (2006) The green gardener: sustainable gardening in your own backyard. Penguin Group, Camberwell, Victoria.

Elliot R (2003) *Australian Plants for Mediterranean Climate Gardens*. Rosenburg Publishing, Dural, NSW.

Eustace RC and Johnston LM (1996) *Indigenous Gardening – Growing Local Native Plants*. McBenny, Cannon Hill, Queensland.

Lullfitz G (2002) *A New Image for Western Australian Plants*. Lullfitz Nursery, Wanneroo, Western Australia.

Mason J (2007) *Growing Australian natives*. 2nd edition. Simon and Schuster (Australia), Pymble, NSW.

Seddon G (2006) *The old country: Australian landscapes, plants and people*. Cambridge University Press, Melbourne, Victoria.

Stewart A (2010) *Creating an Australian Garden*. Allen and Unwin, Crows Nest, NSW.

Urquhart P (1999) *The new native garden – designing with Australian plants*. Lansdowne Publishing, Sydney, NSW.

Wrigley JW and Fagg M (2007) *Australian Native Plants – Cultivation, Use in Landscaping and Propagation*. New Holland Publishers (Australia), Chatswood, NSW.

Chapter 1: Growing Australian native plants

Elliot R (2003) *Australian Plants for Mediterranean Climate Gardens*. Rosenburg Publishing, Dural, NSW.

Handreck K (2001) *Gardening Down-Under – A guide to healthier soils and plants*. 2nd edition. Landlinks Press, Collingwood, Victoria.

Lewis WJ and Alexander DMcE (2008) *Grafting and Budding: A Practical Guide for Fruit and Nut Plants and Ornamentals*. 2nd edition. Landlinks Press, Collingwood, Victoria.

Mason J (2007) *Growing Australian natives*. 2nd edition. Simon and Schuster (Australia), Pymble, NSW.

Stewart A (2010) *Creating an Australian Garden*. Allen and Unwin, Crows Nest, NSW.

Wrigley JW and Fagg M (2007) *Australian Native Plants – Cultivation, Use in Landscaping and Propagation*. New Holland Publishers (Australia), Chatswood, NSW.

Chapter 2: Groundcovers and shrubs

Cavanagh A and Pieroni M (2006) *The Dryandra Book*. Australian Plants Society (SGAP Victoria) Inc. and Wildflower Society of Western Australia Inc. in association with CSIRO Publishing, Melbourne.

Collins K, Collins K and George A (2008) *Banksias*. CSIRO Publishing, Melbourne.

George E (2002) *Verticordia: The Turner of Hearts*. UWA Press, Perth, Western Australia.

Hopper SD (1993) *Kangaroo Paws and Catspaws: A natural history and field guide*. Department of Conservation and Land Management, Como, Western Australia.

Moore P (2005) *A Guide to Plants of Inland Australia*. New Holland Publishers (Australia), Sydney, NSW.

Olde P and Marriott N (1994) *The Grevillea Book*. Volume 1. Kangaroo Press, Kenthurst, NSW.

Olde P and Marriott N (1995) *The Grevillea Book*. Volumes 2 and 3. Kangaroo Press, Kenthurst, NSW.

Parry N and Jones J (2009) *Small Native Plants for Australian Gardens.* New Holland Publishers (Australia), Sydney, NSW.

Stewart A (2010) *Creating an Australian Garden.* Allen and Unwin, Crows Nest, NSW.

Urquhart P (1999) *The new native garden – designing with Australian plants.* Lansdowne Publishing, Sydney, NSW.

Wrigley J and Fagg M (1989) *Banksias, Waratahs and Grevilleas.* Collins Publishers Australia.

Wrigley J and Fagg M (2002) *Starting out with natives: easy-to-grow plants for your area.* Reed New Holland, Sydney, NSW.

Chapter 3: Small and medium trees

Brooker MIH and Kleinig DA (2006) *Field guide to eucalypts Volume 1: south-eastern Australia.* 3rd edition. Bloomings Books, Melbourne, Victoria.

Brooker MIH and Kleinig DA (2011) *Field guide to eucalypts Volume 2: south western and southern Australia.* 3rd edition, Bloomings Books, Melbourne, Victoria.

Brooker MIH and Kleinig DA (2004) *Field guide to eucalypts Volume 3: northern Australia.* 2nd edition, Bloomings Books, Melbourne, Victoria.

Collins K, Collins K and George A (2008) *Banksias.* CSIRO Publishing, Collingwood, Victoria.

Moore P (2005) *A Guide to Plants of Inland Australia.* New Holland Publishers (Australia), Sydney, NSW.

Urquhart P (1999) *The new native garden – designing with Australian plants.* Lansdowne Publishing, Sydney, NSW.

Wrigley J and Fagg M (1989) *Banksias, Waratahs and Grevilleas.* Collins Publishers Australia.

Wrigley J and Fagg M (2002) *Starting out with natives: easy-to-grow plants for your area.* Reed New Holland, Sydney, NSW.

Young JA (2006) *Hakeas of Western Australia: A field and identification guide.*

Chapter 4: Arboriculture

Gilman EF (2011) *An Illustrated Guide to Pruning.* 3rd Edition. Delmar Cengage Learning, Clifton Park, New York.

Harris RW, Clark JR and Matheny NP (2004) *Arboriculture: Integrated Management of Landscape Trees, Shrubs and Vines.* 4th Edition. Pearson Education Inc, New Jersey.

Urban J (2008) *Up By Roots.* International Society of Arboriculture, Champaign, Illinois.

Chapter 5: Propagation

Garner RJ (2003) *The Grafter's Handbook.* Octopus Publishing Group, London.

Hartmann HT, Kester DE, Davies FT and Geneve R (2010) *Hartmann and Kester's Plant Propagation: Principles and Practices.* 8th edition. Prentice Hall, Upper Saddle River, New Jersey.

Lewis WJ and Alexander DMcE (2008) *Grafting and Budding - A Practical Guide for Fruit and Nut Plants and Ornamentals.* 2nd edition. Landlinks Press, Collingwood, Victoria.

Mason J (2004) *Nursery Management.* 2nd edition. Landlinks Press, Collingwood, Victoria.

Wrigley JW and Fagg M (2007) *Australian Native Plants – Cultivation, Use in Landscaping and Propagation.* New Holland Publishers (Australia), Chatswood, NSW.

Chapter 6: Seeds

Baskin C and Baskin J (1998) *Seeds: Ecology, Biogeography, and Evolution of Dormancy and Germination*. Academic Press, New York.

Kessler R and Stuppy W (2006) *Seeds: Time Capsules of Life*. Papadakis, London.

Offord CA and Meagher PF (eds) (2009) *Plant Germplasm Conservation in Australia: Strategies and Guidelines for Developing, Managing and Utilising ex situ Collections*. Australian Network for Plant Conservation Inc., Canberra.

Sweedman L and Merritt D (eds) (2006) *Australian seeds: a guide to their collection, identification and biology*. CSIRO Publishing, Collingwood, Victoria.

Chapter 7: Tissue culture and cryopreservation

Bhojwani SS and Razdan MK (1996) *Plant Tissue Culture: Theory and Practice*. Elsevier Science BV, Amsterdam.

George EF, Hall MA and De Klerk G-J (Eds.) (2010) *Plant Propagation by Tissue Culture: Volume 1. The Background*. 3rd edition. Exegetics, Basingstoke.

Smith RH (2012) *Plant Tissue Culture: Techniques and Experiments*. 3rd edition. Academic Press, London.

Sathyanarayana BN and Verghese DB (2007) *Plant Tissue Culture: Practices and New Experimental Protocols*. I. K. International Pvt Ltd, New Delhi.

Chapter 8: Pests, diseases, disorders and other problems

Common IFB (1990) *Moths of Australia*. Melbourne University Press, Melbourne.

CSIRO (1991) *The insects of Australia: a textbook for students and research workers*. Volumes 1 & 2. Melbourne University Press, Melbourne.

Dell B, Malajczuk N and Grove TS (2001) *Nutrient disorders in plantation eucalypts*. Australian Centre for International Agricultural Research, Canberra.

Jones D and Elliot R (1995) *Pests, diseases and ailments of Australian plants*. Lothian Books, Melbourne.

Keane PJ, Kile GA, Podger FD and Brown BN (2000) *Diseases and Pathogens of Eucalypts*. CSIRO Publishing Melbourne.

Lawrence JF and Britten EB (1994) *Australian Beetles*. Melbourne University Press. Melbourne.

Chapter 9: The selection and breeding of Australian plants

Allard RW (1999) *Principles of Plant Breeding*. 2nd edition. John Wiley and Sons, New York.

Hopper SD (1993) *Kangaroo Paws and Catspaws – A Natural History and Field Guide*. Department of Conservation and Land Management, Como, Western Australia.

Kearns CA and Inouye DW (1993) *Techniques for Pollination Biologists*. University Press of Colorado, Niwot, Colorado.

Olde P and Marriott N (1994) *The Grevillea Book, Vol 1*. Kangaroo Press, Kenthurst, New South Wales.

Olde P and Marriott N (1995) *The Grevillea Book, Vols. 2 & 3*. Kangaroo Press, Kenthurst, New South Wales.

McMaugh J (2000) *What garden pest or disease is that? Organic and Chemical Solutions for Every Garden Problem*. New Holland Publishers, Australia.

Index

www.ingramcontent.com/pod-product-compliance
Lightning Source LLC
LaVergne TN
LVHW070124110826
845147LV00002B/181

* 9 7 8 0 6 4 3 1 0 3 2 1 4 *